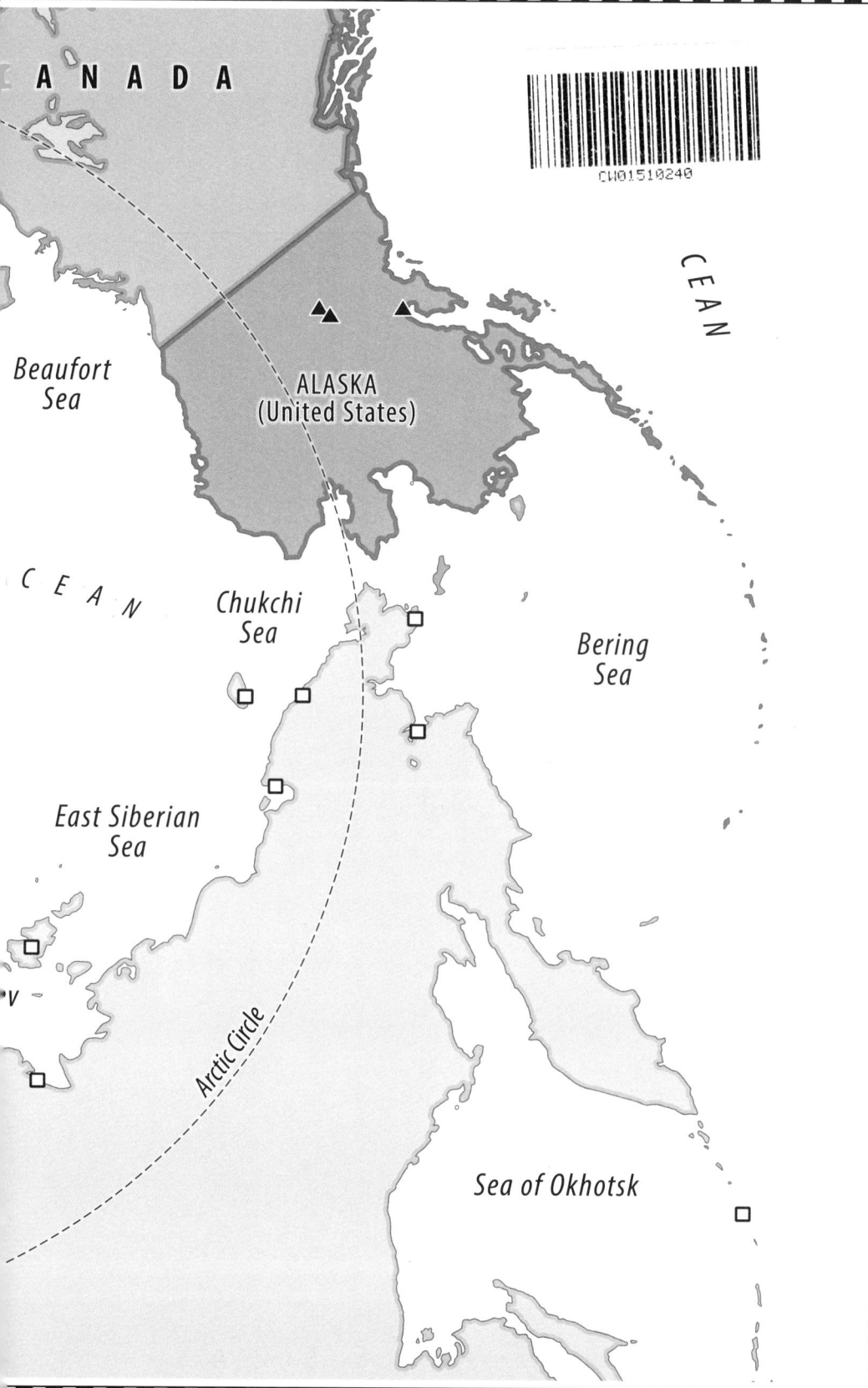

CANADA

CW01510240

Beaufort
Sea

ALASKA
(United States)

C E A N

C E A N

Chukchi
Sea

Bering
Sea

East Siberian
Sea

Arctic Circle

Sea of Okhotsk

POLAR WAR

POLAR WAR

WAR

Submarines, Spies and the
Struggle for Power in a Melting Arctic

KENNETH R. ROSEN

Profile Books

First published in Great Britain in 2026 by
Profile Books Ltd
29 Cloth Fair
London
EC1A 7JQ

www.profilebooks.com

First published in the United States of America in 2026 by Simon & Schuster

Interior Design by Wendy Blum

10 9 8 7 6 5 4

Printed and bound in Great Britain by
CPI Group (UK) Ltd, Croydon, CR0 4YY

A CIP catalogue record for this book is available from the British Library.

Our product safety representative in the EU is BGC Sustainability & Compliance, 7 avenue du Général Leclerc, Paris, 75014, France https://baldwinglobalconsulting.com

ISBN 978 1 80522 912 4
eISBN 978 1 80522 914 8

MIX
Paper | Supporting
responsible forestry
FSC
www.fsc.org FSC® C013604

To the people of the north,
from whom we have taken much and granted little.

And to my family,
for their patience and support no matter the cost.

For days the igloo has been dark;
But now the rag wick sends a spark
That glitters in the icy air,
And wakes frost sapphires everywhere;
Bright, bitter flames, that adder-like
Dart here and there, yet fear to strike
The gruesome gloom wherein they lie,
My comrades, oh, so keen to die!
And I, the last—well, here I wait
The clock to strike the hour of eight. . . .

Robert Service,
"Death in the Arctic"

History is a child building a sand-castle by the sea.

Heraclitus

CONTENTS

POLAR
WAR

A NOTE TO READERS

I WROTE THE FIRST sentences of this book inside a windowless berth aboard a U.S. Coast Guard National Security Cutter in the Bering Sea. I wrote the last passage during a bout with chilblains in arctic conditions far below the Arctic Circle, with the aurora borealis breaching the skies over New England in an extraordinary visit of the northern lights, a time during which a new American president signaled a desire to claim arctic territory from two allies using rhetoric that mimicked the region's greatest historic and living autocrats. A time when additional threats to the arctic were unnecessary. A time of great upheaval.

One could be forgiven for thinking tensions in the arctic, perhaps most prominently embodied by President Donald J. Trump's egregious campaign to "get" and secure Greenland as another American territory, came out of nowhere. But it did not. Historically, the American desire to control Greenland has existed nearly as long as America itself. It was not Trump's rhetoric of a takeover that struck me during my years spent traveling and reporting on the circumpolar north. It was the ineptitude surrounding the idea. To publicly make threats of invading Greenland while America continually fails by all metrics in the arctic, at home and abroad, seemed anathema to our own desires and lacking abilities across the region. This book therefore comes at a time when there are more questions than answers for the future of the American and global arctics alike. Its modest aim is to more fully

1

comprehend what some have called a new Cold War, its history, its future, and what's at stake for everyone, whether they reside below or above that invisible circle separating snow and ice from the rest of the world. To those of us who have been watching the arctic over the last half-decade or so, the possibility of conflict in the arctic now feels inevitable.

One premise of this book is that America has for too long neglected its arctic territory—primarily Alaska—and its broader commitment to cooperation and solidarity across the high north as the region grew hotter, more contentious, less stable. In the last year, however, a reversal of sorts occurred. A new administration placed a premium on the north, a place wherein it saw the potential for ever more exploitation and militarization, echoing a hallmark strategy of the Russian Federation. President Trump, in August 2025, used a military base in Anchorage as the backdrop for a summit with Russian President Vladimir V. Putin, exposing the state for what it had become to Trump's administration: a springboard for matters of war.

We have refused to learn from the follies of the past, when the heydays of Alaska led to rushes for gold and gas and the segregation of native tribes from pristine stretches of wilderness, and this approach will prove fatal for industry, for national security, and for the people who reside across the American Arctic.

The circumpolar north will always remain important for the same reasons it stays with those who find themselves susceptible to its otherworldly influence, viscerally clinging to those who accept its omniscience no matter one's distance from the geographic North Pole, and especially no matter their politics. In the way Massachusetts found itself kissed by the aurora, the subject of the arctic, its permanence atop our world, seems to return with greater frequency each day, reminding us that we are all but children of the sun.

Any and all errors are mine and are regrettable.

K. R. R., September 2025

FLAGS

YEARS ACCRUE OVER TWO seasons, never four. Dark days and bright nights are shared experiences in the circumpolar north, the arctic, where time is accounted differently, the midnight sun a warning flare slung into the sky before disappearing into darkness each and every winter. Darkness, then light, then darkness once more, a land of infinite extremes. More than this, though, one is struck by the elastic distances, landscapes over which the radical whims and decisions—of individuals, of nations—vacillate. Geopolitical instability like that occurring in Greenland, or in the North Atlantic, or on the isolated Svalbard archipelago north of mainland Norway and reverberating elsewhere does not necessarily lead to war, but it has happened before, even in a place as strange as this, even in a place where any war might seem improbable. After all, wars take many shapes.

Yawning below the tiny twelve-seater, a histogram of white, an eerie topographic menace. Jagged shapes, shadowy splinters, and ice claws with no end in sight. It is midday in December, in the cockpit of a Dash 8 somewhere over western Greenland, and only a thin line of pale red, evidence of the sun to the south, segments sky from earth. We pass below 75 degrees north and the digital compass display, dormant for the past hour, blinks to life as the pilot switches from manual to computer-assisted navigation.

The experience of transiting the arctic's vastness has changed dramatically. Once it took courage; today it requires only time and cash. Sparsely populated and not permanently settled until recently, the ice and snow and formidable cold below were once a proving ground. Men and women on missions to chart uncharted land were met not by hostile Indigenous tribes, but by beasts of weather and fur. Some foolishly wanted fame, others glory for their nation, still more were merely curious about what lay north. They were tested. Many failed. Now, to get from Greenland's northernmost village to its southernmost enclave, a trip of some 990 miles over ice and snow and glaciers in various phases of collapse, takes a few hours by plane; the same historically spectacular views, none of the agony.

The pilot reaches for a knob to recalibrate our position. From the flight deck, he and his co-pilot had been navigating by true north until this moment, never magnetic north at those latitudes. So close to the Magnetic North Pole, and at a lower altitude, the readings cannot be trusted. I shudder at the blinking lights, of what I imagine to be the failure of the plane's directional bearing. I later speak to Claus Kongsgaard Jensen, who flies the same route. He tells me with the confidence of a military veteran, "You just reset it." Jensen, by his own count, had flown 50,000 hours across the arctic. He joined the Royal Danish Air Force and began his tour in Greenland in 1988. He was also a veteran of a little-known arctic skirmish known as the Whisky War. Because of that war, and those that would inevitably follow, he knew intimately that the arctic, like the flight deck's compass, was always in a state of geopolitical and ecological recalibration.

The Whisky War, if you find utility in the kind of declaratives and certainties that the arctic is never sure to grant, began during the polar expedition of Charles Francis Hall in 1871, when he discovered a spit of land—a large rock, really, slightly longer than a half mile—somewhere north of Baffin Bay and south of the Lincoln Sea, between the northern reaches of western Greenland and Canada's Nunavut territory on a tassel of water stretching into the Arctic Ocean. Hall named the island Hans Island after Hans Hendrik, a Greenlandic guide with the expedition who had skirted the island two decades before with the American explorer Elisha Kent Kane. More than a century and two world wars later, the border between Canada and

Greenland was officially drawn in 1973, and the island fell once more into the international spotlight. The low-tide levels on the northern and southern edges of Hans Island marked where one nation ended and the other began, a boulder snug between them.

The matter seemed settled. A provision in this border agreement allowed for future negotiations over the entire island, not just its circumference. Then, a decade later, a Canadian scientist from Dome Petroleum arrived on Hans Island in 1983 to research the impact of multi-year ice in an effort to learn how best to construct oil-drilling rigs in the Beaufort Sea. Learning of this expedition and believing it to be an assertion of Canadian sovereignty over the rock, Tom Høyem, the Kingdom of Denmark's Minister of Greenlandic Affairs, set out on his own mission to assert his country's sovereignty over the rock and the adjacent island nation it claims as its own.

Høyem chartered a helicopter from an American military base outside Qaanaaq, 217 miles south of Hans Island, and arrived on the island in 1984 to deposit a Danish flag, a bottle of aquavit, and a note reading: "Welcome to the Danish Island." Soon the Canadians arrived and replaced the Dannebrog flag with a Maple Leaf flag, the aquavit with Canadian Club whisky. The bottle and flag swapping threatened to become the battle of the bottles. The fusillade carried on for nearly four decades, bottles for bottles, flags for flags. Never mind both nations' territorial claims were spurious at best. Canadian Indigenous people never used the island; no residents of Denmark's realm— including Greenland and the Faroe Islands—had ever permanently resided on the island known as Hans. The rock, as with many things in the arctic, stood for something much more. Neither country wanted to appear like it was yielding their stances on arctic sovereignty. But saber rattling and chest thumping were the instruments of those seeking assertion in the arctic, and besides, at least a few transient residents in the region benefited as the liquor bottles went to good use. At least one Danish pilot returned to Hans Island to snatch the bottles and later drink them among friends at the American air base on Greenland's western shores.

Airmen like Jensen battled over this insignificant and practically useless swath of territory in the north throughout the Cold War, this political spat eclipsed by nuclear threats and missile crises beyond the Iron Curtain. Nevertheless, Jensen flew often within the Arctic Circle while defending that rock, playing a small role in the war fought upon and in the skies above Hans Island. There were no trees, no streams, no landmarks of any kind, not a trace of civilization there save for the flags and bottles. There were no untapped natural resources, not a thing beyond the niggling tenterhooks of national pride. Yet the campaign ensued religiously for nearly fifty years. The arctic has this effect. It is called polar madness.

There was perhaps at one time a reason for the island's significance to both nations. The Canadians operated a scientific station on Hans Island during the Second World War, ostensibly to gather weather data in support of a secret Allied invasion of France, and the Canadian military occasionally conducted exercises in the region around the island. Though it possesses no usable mineral reserves, and though it is barren and cold and surrounded by impossible currents, Hans Island might someday be an ideal staging ground for offshore oil drilling as the arctic sea ice retreats and disappears more every year. Which is also a trope of the arctic and the war: the very ice the Canadian scientist from Dome Petroleum had first come to study, the research which propelled two nations into conflict, would soon thereafter vanish. In its place were uncertainty, a reexamination of purpose, visions, and both commercial and political calculations.

To finally decide who was entitled to Hans Island, politics ensued in a region where politicking is like nature itself: enduring and untamable. It is here that settling such territorial or inter-governmental disputes requires no less than a joint working group, representatives from Inuit communities in Greenland and Canada's Nunavut region, environmental organizations, and a wealth of speculations feeding excitable press coverage. The process of deliberation and settlement took roughly five years. The final agreement, reached on June 14, 2022, was celebrated with one last exchange of liquor bottles, resolving a dispute over an island the Inuits unironically call Tartupaluk, or "kidney," for its shape and not for its mess of liquor bottles. Though the Whisky War may seem absurd, it would be difficult to downplay the undertones of the conflict that persist in the context of today's changing arctic, a region where Russia frequently exerts pressure on North Atlantic Treaty Organization (NATO) allies' critical infrastructure, like submarine telecommunication cables. As Rear Admiral Rune Andersen, the commander of the Norwegian Navy, told me, "What happens here impacts everywhere."

THERE ARE FEW THINGS held in the popular imagination beyond polar bears, snow, ice, and "Eskimos" when calling to mind the arctic. Anyone who has lived in Alaska, as I once did, or anywhere in the far north, knows that there is much more than that when it comes to the vast circumpolar cosmos crowning the Northern Hemisphere.

Each year the planet permanently loses a section of sea ice the size of Nebraska, along with it the reflective properties of that snow and ice, which will no longer return heat to the atmosphere. Every year the Arctic Ocean sees new summer minimum-ice-extent records, advancing a downward trend of 12.5 percent per decade since 1979. When the Arctic Ocean sea ice, which one can imagine as a gyrating glacier at the top of the world, melts, this does not increase sea levels any more than a melting ice cube would cause your drink to overflow. (Greenland's ice cap is another story.) Yet as the ice melts, more of the ocean is exposed, and more exposed deep blue water acts as a sink into which the excess heat of the sun swirls. The warmer water sets in motion what scientists call a feedback loop. The excess heat dries up coral reefs, alters marine chemistry, and can upend thermohaline circulation, the process by which salt and water temperatures are dispersed throughout the world's oceans.

This circulation, long predictable and reliable, shapes the earth's climate. This circulation, when mixed with wind-driven surface currents, is what makes for milder temperatures across Europe relative to the temperatures at the same latitudes in North America. It was what once allowed us in the Northern Hemisphere to safely assume that in September we will need to change our clothes to something warmer, that in late March we once could expect to start doffing our winter layers. But not today. Summers last longer, winters are shorter. A slowing of this circulation, which is sometimes known as the ocean conveyor belt, could have unforeseen consequences for our climate. An increase in warm water from the sun's heat is just as challenging to the health of the planet as an increase of cold water from the melting of glaciers. The last time a rapid influx of cold water was introduced into that conveyor flowing between the world's five oceans, global temperatures plunged humanity into an ice age.

Today, similar changes impact animal and human life equally along the shores of the Arctic Ocean. Marine life is being displaced and its biology is changing; humans who rely on that aquatic life are finding those species less accessible. Meanwhile, the heat disrupts the foundations of their homes, thawing the permafrost on which they sit. It changes the physical landscape and how those situated upon it live, work, and play. The most important biosphere known to humanity, from its greatest predator (*Homo sapiens*) to its smallest, was once an icebox. Now it is an oven.

Nations near and far from the region have seen the warming as an opportunity for expansion and military dominance, and what they have cooked up is a conflict teetering toward full-blown war, even in a place that has largely been historically regarded as inaccessible to geopolitical ambitions. North American and European officials no longer see the arctic as a "zone of peace," a term coined by Soviet leader Mikhail S. Gorbachev, but rather one of war. This was impossible to see when I lived in Juneau, the capital of Alaska, for what turned into several months of hysteria in 2013. The first thing anyone in the arctic offered me then was a stiff drink. The second was a gun. The broader region was like this—a welcoming place, *in extremis,* where first you are welcomed and then, rather immediately, alerted to its dangers.

For decades during and after the Cold War, and still today, the complex dovetailing of national interests and disinterests, paired now with the rapid changes to the climate across the arctic, have become more dangerous, and seeing them play out can be as jarring as watching a plane's compass go dark. At times comedic and at times morose, conflicts like the Whisky War have spurred nations into fighting over, dividing, marking, demarcating, longing for, and pummeling themselves into indifference over their arctic ambitions. There have been serious conflicts and near-conflicts (as when an American nuclear submarine collided with a Russian nuclear submarine off the northern coast of Russia) alongside more innocuous and ridiculous episodes. What has changed in recent years is their frequency and their consequences for those living inside and outside the arctic.

Russian spies. Nuclear submarines. Declarations by Asian countries claiming they are "near-arctic" nations. Undersea communications severed at night. Sabotaged gas pipelines. Apartment buildings, hospitals, and homes crumbling daily or being wiped away by rising seas. The arctic regions today are home to spycraft and nuclear testing, in addition to cruise ships and booming commercial shipping lanes.

Seen through the lens of expanded and cheaper free trade as a boon to commercial investment, the melting of ice caps no longer ushers in well-meaning scientific collaboration but, rather, fighting for new trade routes. A new arctic route would shorten by two weeks the distance traveled by tankers and cargo and passenger ships from East Asia to Western Europe, circumventing the Suez Canal, across a route largely dominated by Russia. This would be good news for consumers and disastrous for the climate as the northern world barrels toward a Cold War reborn.

One of the silent characteristics of the last Cold War was the way Eastern and Western superpowers transplanted their interests onto foreign terrain, ignoring the regions and nations regarded as subordinate and which lay in opposition to their aims. The Cold War became a clash of cultures and traditions marked by exceeding arrogance for lands apart from those superpowers' own. Throughout the Cold War, the United States put a lot of stake in the arctic, understanding it as a breadbasket of sorts. It could supply sensational natural resources and defensive outposts. Military contractors and scientists wrote reams of studies about Alaska's significance to the nation's maritime industries and defense, taking great interest in the finer points of ice science, the cryozone, cold-weather construction, the health of both the planet and this fragile region. After the fall of the Soviet Union, tensions cooled, and the north was once again a place for exploration and excellence on the world stage. It was an arena of partnership and cooperation, where science diplomacy, joint search-and-rescue operations, wildlife preservation, restitutions for Indigenous people, and natural-resource extraction were beneficial to all nations with northern territory, even those without.

"The community and interrelationship of the interests of our entire world is felt in the northern part of the globe, in the arctic, perhaps more than anywhere else," Gorbachev said in 1987 while in Murmansk, on the Kola Peninsula above the Arctic Circle, today still the location of the bulwark of Russia's nuclear weapons. The Soviet Union, Gorbachev said, had a "profound and certain interest in preventing the North of the planet, its Polar and sub-Polar regions and all Northern countries from ever again becoming an arena of war, and in forming there a genuine zone of peace and fruitful cooperation."

But what even is the arctic exactly? There's little dispute that the arctic is a dark—literal, and figurative—place. But its definition is precisely illusory and even the exact latitude at which the region begins is not agreed upon. The arctic may be divided by latitude into two regions: the arctic proper (population 4 million) and the subarctic (population 13 million). This offers a loose definition of the *arctic* (there is no formal definition), but it's far from the only one. The people who live, work, and study there define it differently and for different reasons, resulting in definitions as varied as the land and seas across the circumpolar north.

Some choose to define the arctic as only anything above the Arctic Circle at 66 degrees and 33 minutes north, delineating where you may see the midnight sun during the summer solstice, and twenty-four hours of darkness during the winter solstice. Some may say the arctic begins where the taiga (Russian for "land of the little sticks" and the world's largest land biome, also known as the boreal forest) ends and where steppes become tundra. But the boreal's boundaries are moving. Smaller-growth trees creak and creep and crawl northward as the planet warms.

Others may say the arctic begins where there is permafrost, that layer of frozen substrate. Others still may point to median annual, monthly, or daily temperatures. Yet the planet tilts irregularly and impedes the precision of solstices and equinoxes, permafrost is spotty even in areas of the high north, and average temperatures swing like a pendulum. Thermometers across the Canadian Arctic might not reach below minus 51 degrees Fahrenheit, while in Siberia lows of minus 80 degrees are typical. Each arctic nation has its own definition of their territory's boundaries; Iceland itself is an arctic state, despite being a small island just shy of the Arctic Circle. The word *arctic* derives from the Greek word for "bear," and the name *Antarctica* reminds us that no bears inhabit that southern continent. When the last of the polar bears die, then I suppose the name *arctic* will merely become an honorific—in memorial, to a world in perpetual flux.

The uncertainty and constant changes have never been more consequential for homeland security: an increase in militarized fishing vessels, nuclear weapons tests, and a reallocation of state funds from regional community-oriented enterprises to natural-resource extraction and defense spending. Military buildup is erasing communities while refueling a race for power dominance in the north, advancing regional tensions that until now were historically subdued. With the climate warming, it seems we have rather silently entered a consequential race for dominance in a region long held as a place of provenance and collaboration.

At the 2021 Defense Threat Reduction Agency's Strategic Forum, Jill M. Brandenberger, who was the climate security program manager at Pacific Northwest National Laboratory, told a modest crowd of defense and security professionals that increased access, foreign direct investment, and technology innovation were three "mega threads" across the arctic because of climate change. Dr. Brandenberger, a coastal oceanographer, made clear in her forty-five–minute presentation that a storm of strategic jostling was on the horizon. "The convergence of these three mega threads are [*sic*] really creating specific competition areas . . . in particular between Russia and China that we, in the defense and national security industry, need to follow."

WAR IN THE NORTH is wildly unthinkable. Most will say that such a conflict is improbable. It is too cold, too harsh, too inaccessible, too unpredictable. The cost-benefit analysis of conflict here does not add up. What's more likely, I've heard time and again, is that a maritime emergency—a stranded cruise ship, a botched fisheries inspection that ends in a gunfight between the nation being boarded and the nation doing the boarding—will lead to conflict elsewhere, a military phenomenon known as horizontal escalation. Indeed, classified tabletop exercises to prepare officials and those likeliest to respond to these incidents, to gauge preparedness and the potential operational capacities of military and their affiliated government bodies, have become a hallmark of broader, more frequent discussions in Washington about the arctic. This belief that a war would commence only by accident conceals fundamental advantages of military and political operations in the arctic. As Mark Rutte, the NATO Secretary General, suggested in July 2025, nations unfriendly to the alliance could distract from a larger regional conflict elsewhere in, say, Taiwan, by attempting to stake claim to NATO arctic territory. Aside from the diplomatic advantages of maintaining closer ties with northern allies at outposts across these regions, a military can deploy forces quickly by transiting across the North Pole. And if a military is practiced in cold-weather warfare, it can operate virtually anywhere. Militaries have learned over the centuries that a war in the arctic is advantageous only to those who have mastered it.

Viewed through this prism, unity was always a dark horse in the north. In 2007, two twenty-six-foot-long Russian submersibles descended from the ice at the North Pole and planted a small Russian tricolor on the ocean floor. Onboard the submersibles were a Swedish businessman, a sheikh from the United Arab Emirates, an Australian adventurer, and the vice-speaker of Russia's cosmetic parliament, the Duma. When the explorers came home, they were welcomed like astronauts returning from a trip to a new planet.

The reception elsewhere was cooler. "This isn't the fifteenth century," Peter MacKay, Canada's foreign minister, said at the time. "You can't go around the world and just plant flags." The Russian president, feted as a *Time* Person of the Year that same year, tried to diminish international worry. If anything, it was playful brinkmanship, like the Whisky War. "Don't worry," Russian President Vladimir V. Putin said. "Everything will be all right."

SOME YEARS AFTER HE stopped flying reconnaissance missions across Greenland, to and from Hans Island, Jensen wrote his master's thesis, at the Royal Danish Air Force Academy, on "The Battle of the North Pole." "My teachers at the academy called it an 'irrelevant discussion,'" Jensen told me. "They did not believe there would be any discussion about the North Pole in a military sense. Everything would just be by diplomatic means, like Hans Island." What Hans Island teaches us, then, after considering the broader history of ambitions and disagreements across the northern regions, is this: Conflicts in the arctic are resolved first by threats of military intervention and then settled, to the detriment of regional stability, after decades of passive reflection and harm wrought on communities and wildlife, with the naive expectation that none of it was necessary. It also exemplifies the most common type of conflict in the arctic. The Whisky War may have ended, but the juvenile "attacks" and the imprudent gamesmanship and antagonism are not missing from the arctic today. Russia upgrades a base on the shore of Kotelny Island in Siberia, and the United States relocates fifth-generation fighter jets to Alaska. China sends a research balloon soaring over the Canadian Arctic, and the North American Aerospace Defense Command (NORAD) shoots it down. One nation sends two submarines under the North Pole; another nation sends three.

In the years since the Hans Island settlement, flags and territorial claims continue to play their role in a region of evermore complexity and competition. In June 2024, the Russian Federation returned to the North Pole. A rather innocuous display of national sovereignty—a flag—was this time replaced by one bearing the letter *Z*. In the years since Russia's renewed insurgency in Ukraine, that letter has come to stand for much to many people: fascism, indiscriminate bombing and killing of civilians, the kidnapping and displacement of thousands of children. Now the letter and all it represented had arrived in the north, bearing the markings of a new Cold War, a horizontal conflict, a hastening of preparations for military confrontation. The directional compass of the region's heading was again being reset.

Jensen's thesis, it turns out, was prophetic. Discussions of the north's militarization and potential as a sphere for conflict or competition have grown louder at military academies around the world, perhaps most of all in America. Of the five military academies nationwide—West Point, the Naval Academy, the Air Force Academy, the Coast Guard Academy, and the Merchant Marine Academy—along with a handful of U.S. service academies like the U.S. Marine Corps University, not one is today without a course on the arctic. Only a decade ago such classes would have seemed like gratuitous extracurriculars. Many now offer several courses on the region, as enrollment and interest in the region soars and military leaders seek to establish within the minds of service members, long focused on conflicts in the Middle East, Southeast Asia, the Pacific theater, the Global War on Terror, and quick-reaction expeditionary forces, that the arctic is no longer a sleepy zone of peace. One resident of the European Arctic told me the arctic had become anything but that. For all their years spent living and working there, the arctic was becoming, had become, the exact opposite of Gorbachev's sentiment. For what are flags if not declarations? They are used at times to possess and repossess, to unify and divide, and eventually, to conquer. "I don't want to be this alarmist person to say the war is coming," they told me in confidence, "but the war is fucking coming."

CLOUDBERRIES

THE WAR MY FRIEND envisioned on the horizon was easily seen. Then again, in my mind, the arctic had always existed as a vast region that was, formal definitions be damned, a loose arena of depravity and sacrifice and conflict—a region not of auroras and dogsledding, of ice hotels and whale watching cruises, but of barroom brawls, substance abuse, and temperatures to match the disposition of the already dispossessed. The region has in fact always been viewed through caustic extremes.

Ancient mythology held that this land beyond the North Winds was a temperate climate, a *terra incognita* where the gods hovered within reach. Hyperborea, as it was called in Greek, was supposedly guarded by harsh cliffs of ice and rock. Here it was said white swans alighted on poplar trees dipped in amber. The sun chased the moon, neither one ever settling for less than complete exposure and ushering double grain harvests in a disease-free land of the impossible. Here was utopia. Here was the seafarer's Great White Whale to be caught, tamed, and slaughtered. Here lived a strong, undeterred people. The Roman writers Virgil and Cicero said the Hyperboreans lived for a thousand years in complete euphoria, masters of their universe beyond what were believed to be the Riphean Mountains, a natural bastion providing the tyranny of scale and distance from southernly worlds and beyond. Ancient writers believed Hyperborea to be here in the north. No one had ever truly known, or had seen it for themselves.

For centuries these myths, or variations on them, about the northern

15

circumpolar region persisted in the proclamations of poets, historians, writers, geographers, explorers, and statesmen who sought—for want of adventure, of claims to have fleeced the exotic—an ancillary, inconsequential, ice-bound, uninhabitable wasteland with little direct import for temperate nation-states and climates that hosted the bulk of the world's civilization. It seemed they merely wanted to conquer.

The Phoenicians and Greeks knew of a place they called Thule, "though most probably" they had never reached what Naddoðr the Norwegian Viking reached in the ninth century: what we know today as Iceland. And while it seemed that the western—or in this case southerly—world had finally touched the barren, forgotten region (evidence of early Irish Christianity in Iceland is perhaps a testament to the reach and voyages of the sixth-century priest Saint Brendan of Birr), the arctic was by then already a thriving home to humankind, inhospitable only to the unappreciative.

What was unknown about the land and those people living across it made excellent, if overexcited and exaggerated, fodder for the literary public. In 1555, Olaus Magnus wrote in *A Description of the Northern Peoples* that those same people lived much longer than their southern contemporaries, and that they were tormented by sea monsters. In his fantastical *Islandia*, first published in 1607, Dithmar Blefken claimed people of the north lived for 150 years. Pierre Martin de la Martinière, the French explorer, wrote in 1671 in *Voyage des pais septentrionaux* (The Northern Lights Route) that those living above the Arctic Circle were "sorcerers, who control the winds at will." They called them *Esquimaux* (from the French for "raw meat eaters"). By the late nineteenth century, it was clear that despite the deepest silences born and lain to rest in the arctic, "the limits of the unknown had to recede step-by-step before the ever-increasing yearning after light and knowledge of the human mind." More explorers and inquisitors thus set their sights northward.

Today, scientists tell us this is not Hyperborea, but simply the arctic, home to dozens of Indigenous tribes and towering walls of plethoric sedimentary limestone and sandstone deposits atop fanlike deltas where eroded rocks empty into the tributaries of nearby fjords. Humankind came to conquer and know the arctic well, or so it believed. The world's actions later unraveled what had been known about the arctic. Carbon-emitting technologies and fossil fuels trap heat inside the planet's atmosphere, warming the land and the oceans, melting glaciers and ice caps and pushing sea levels higher, inciting typhoons and unpredictable storms, the strength of which are not native to the region. (Since the earth bulges as it spins on its axis, the water levels at the North Pole are roughly thirteen miles lower than they are at the Equator. This can make the search for the meaning of "sea level" quixotic, strange, and effectively meaningless.) Instead of heeding any warnings, adapting humanity's behavior to the land, being mindful of the changes and their consequences, nations have instead begun to act as though a warming planet is an opportunity, not a damnation. Yet, if one stayed long enough to receive the gifts bestowed by the arctic, paradise would eventually look like perdition.

Later narratives belie this point. Those that portray any part of the far north as fixed—euphoric tales of nineteenth-century explorers, or of the contemporary scientists on missions to understand the region—are disingenuous at best. They are simple repetitions glancing on the subject matters of climate change, the extraordinary wonderment of the region, and the history that may be lost because of global warming. They elide the region's responses to those changes, the region's future, and the political mobilization and militarization that have followed.

Though this evidence of a new Cold War has been documented since at least the early 2000s—certainly more since my time in Alaska—the relationship between a warming planet and the region's militarization was barely a footnote in many nations' global ambitions until about 2010 or so. Russia is leading the charge, with more military bases in the arctic, greater competency in cold-weather operations, and a fleet of icebreakers that dwarfs the maritime arctic fleets of every other nation. America and its allies have meanwhile played catch-up. For the United States, whose military was slow to recover an enemy craft because of inexperience in the dark days of the high north and the accompanying frigid blasts during a recent exercise, this involves learning to fortify and base its military units and protect its critical infrastructures like oil pipelines in a climate where it has little modern experience and capability.

Since the Cold War, the U.S. government and military seemed to awaken somewhat to the threat of foreign dominance of the arctic, specifically Russia and China, and to a threat exacerbated by the changing climate. This was followed by the opening of a presence post in northern Norway, and two years later, the appointment of an ambassador-at-large for the arctic region within the State Department, as well as a deputy assistant secretary of defense for Arctic and Global Resilience. America's European allies, too, have been rethinking homeland security, increasing their national defense budgets (in some instances, doubling their spending) and their security around critical energy infrastructures in the arctic, as they aim to boost their defense capabilities—all this with those nations' arctic territories assuming a greater role in NATO's strategy as American assistance and support wanes.

FOR MY PART, I stayed too long and grew mad. During those four long months of hysteria I spent in Juneau, I whittled away much of my time at the Alaskan Hotel and Bar, which occupied a Victorian building on Franklin Street in the capital of Juneau. As far as I know, it still does. That year was the hotel's centennial. The proprietors had opened it in 1913, during the great Alaskan Gold Rush, ceremoniously attaching the front door key to a balloon and setting it aloft, signaling to all that this place, this humble refuge, would be open to anyone, at any time, in perpetuity.

Despite this hospitality, and the warmth that radiated from within, a man who regularly sat there—someone who spent his days operating the tram up to Mount Roberts and his nights along the cracked edges of the Alaskan's worn bar—told me he felt claustrophobic. "There's no horizon," the man said. "We're surrounded, as far and as high as the eye can see, by mountains." I, too, became claustrophobic. So I left.

By the time I return to the arctic in 2022, on a magazine assignment to the region, the world seems to have shifted its focus northward. Behind what I hear described as the "new ice curtain," I meet Timofey Rogozhin—

burly, brownish mop of hair, late forties, square chin, known around town as "the Russian"—in Norway's arctic region, also called the "High North," or the Finnmark region. He lives in a small fishing village about 620 miles north of the Arctic Circle. We had met elsewhere in the arctic on previous occasions, for it is a small, communal region despite its expansiveness. Timofey, echoing the bar patron's sentiment in Juneau, was feeling claustrophobic in a region he had since birth called home. It was a region of open spaces, now slowly becoming one of constriction, tyranny, and war. Timofey was born in Murmansk, a Russian city of about 250,000 and the arctic's largest, on the fringe of the Barents Sea hugging Norway's and Russia's arctic coastlines. After criticizing the government's mistreatment of communities and industries in the Russian Arctic, he fled the country and settled here in Norway.

He speaks very little. He instead takes to social media, castigating the oppressive Russian government, belying his silence in person. He nods to one of the walls in his house and says, "We have a smallish place," and folds modestly into himself. His wife prepares king crab legs and crepes, Russian mushroom canapés, and microwaved finger foods. After dessert (martinis, red wine), a slipper-clad Timofey descends into the cellar and returns holding two jars of preserved cloudberries. The jars look like snow globes, the grains of canning sugar swirling and settling and swirling again as he speaks and gestures with them held tight in his hands. He'd collected the berries from across the basalt highlands, having trekked through the ferns and lands crosshatched by paths trodden under grazing reindeer.

A warming arctic means many things to many people, many nations, many species. Timofey blames the media for exaggerating climate change, believing it to be a scapegoat when the real enemy to the north was Russia, and governments that misunderstood the land he loved. Really, it seems to him, the blame for climate change and the changing arctic lay largely to the south, with the capitals that govern their arctic territories from afar, rarely from within, and that were largely blind to the repercussions of both their actions and their inaction. The intrusions of governments and tourism: Were these afflictions a response to change or were they the change itself? The cloudberries offered some clues and a concise history.

Ripened under the midnight sun, cloudberries are a delicacy ("arctic gold"), selling for $15 per kilogram. They are native to places across the arctic, including Norway's mainland in the north. Across the sea some 900 miles to the north, on the Norwegian archipelago of Svalbard, where Timofey's daughter lives, the thornless berries cannot grow. It is too cold there, the ground too impermeable—that at least was the case until the summer of 2023, when record-breaking heat brought forth the first cloudberries to ever appear there—perhaps the only kind of welcome invasion the arctic will ever witness. Soon, Timofey's daughter will be able to pick and can her own. This was how I return to the arctic: to find it a shadow of its former self, virtually unrecognizable in just a decade's time. And while whispers of a Cold War were growing, I found myself scrambling and shouting to see it firsthand.

For America, the arctic will become strategically exigent. Washington will chair the Arctic Council—the primary forum for political alignment in the northern regions—in 2031–33, overlapping with the fifth International Polar Year. In those years Washington will have much to answer for when it comes to the arctic. Yet here, as with everywhere in the circumpolar north, the steady drumbeats of hardship and change and uncertainty are a preamble to a quiet conflict I once mistakenly believed was frozen—or altogether impossible in a region so harsh and uninviting. Nevertheless, encroachments of all kinds, like the cloudberries, have begun.

Timofey, squiffy from the drinks, thrusts the jars forward.

I stumble and hold them tight, fragile and vulnerable glass vessels I am fearful of crushing, knowing somewhat intuitively that this moment is my own precipice into war.

He has a small favor to ask. He says one jar is for my family. A taste of the arctic, a gift from an upended world. The other is for his daughter. She lives in Longyearbyen. Only if I would not mind.

OUT THE ROAD

A BELL CHIMES OVER the café door in Longyearbyen, a Norwegian town roughly 800 miles south of the North Pole and ringed by the Arctic Ocean. Timofey's daughter shuffles inside. She is tall and wears a woolen cap that frames her face and covers her short dark hair. She takes great strides and reaches the table where I sit. She smiles. I smile. No introductions or exchange of words follow. Those are gestures of little currency in a place of such propinquity. We'd met once before on a previous stay. Then came the hallucinations. Or, did we meet earlier on this trip? It is a small town, a small region. Time is a vacuum. I unzip my bag and reach for the jar. Timofey's daughter cradles the cloudberry jar, smiles, makes a sort of curtsy, then turns, and the bell above the door chimes once more.

Hours earlier, the first thing I notice when arriving at the airport in Longyearbyen on the island of Spitsbergen—a single runway, a single requisite angry stuffed polar bear looming over the single U-shaped baggage carousel inside the standalone terminal—are the expeditionary forces. It is late summer and they are dressed in Gore-Tex and awaiting massive bags, large enough to transport their corpses home were they to die out in the cold. The bags are stuffed with equipment to prevent that from happening, though. They have many straps and buckles. Plenty of webbing. The giddy, trim men and women wear trail-running shoes and stand in a frenetic gaggle. They say they are training for cold-weather survival. They work for a shipping com-

pany preparing its staff for the opening of the Arctic Ocean and its anticipated ice-free summers. They keep to themselves. Guarded. Maybe not a
shipping company. I met one man at the hotel bar in Oslo, between connecting flights (layovers are often counted in days and nights, rarely hours). His
solipsism that evening seemed opposed to the very openness of the place to
which we would soon depart.

Spitsbergen is a decidedly paradoxical island, precariously existing in
a realm somewhere between unarmed conflict and peace. It is a fabulous
introduction to the broader region's historical and modern tensions and
uncertainties.

The larger archipelago, known as Svalbard, was discovered by the Dutch
explorer Willem Barentsz in 1596 and went largely unused and uninhabited
for centuries. In the waning days of the First World War, after "the war to end
all wars" shattered generations and forever changed global borders, Spitsbergen became what some hoped would be an everlasting refuge from conflict.
Fifteen nations—among them the United States, Denmark, France, Russia,
and India—agreed to share the land for commercial, mining, industrial, and
maritime activities, and their citizens were given equal rights to live and
work on the arctic islands. Though overseen by Norway, the archipelago was
exempt from many Norwegian laws and its citizens still need a passport to
visit. It was a demilitarized, visa-free zone. Nowhere else in the world could
one simply arrive and become. A fresh start and the temptation associated
with a life led in the northern backcountries was available only there.

Longyearbyen, the largest of two permanently settled towns on Spitsbergen, is home to around 2,600 year-round residents. More polar bears live in
the region than people. In winter, it needn't be said, the sun does not shine.
This isolated community, the northernmost perennially inhabited on the
planet, is not for those discomfited by distance; it attracts scientists, art-
ists, and hardcore outdoor enthusiasts from all around the world, who live
and work in frozen, impossibly hostile conditions on a collection of islands
over which Norway (its capital, Oslo, is 1,200 miles to the south) still asserts
its sovereignty. For many decades the islands reflected the best of human-

ity, until its resolve was tested, the treaty tried, and arctic cooperation was replaced by competition. Change was happening everywhere here, all at once.

The airport is a drag. Everyone is smiling but few speak; it's as if they are fearful of strangers. A line forms by the doors leading to the arrival hall (which is also the departure hall) as people pause to snap photos with the bear. Many resemble the Michelin Man, were he to dress in an ankle-length Hermès puffer trimmed with synthetic fur. Visit Svalbard, the local tourism organization, posts "How to Dress" videos to its website to help tourists with the finer points of cold-weather layering, but the videos never mention couture. Those who know the arctic do not dress like neon bunnies, preferring simple khakis and several otherwise inconspicuous layers under a thinly insulated jacket. Outside, another line of picture-takers forms by a sign warning of the polar bears and a navigational vane listing the distances in kilometers to various capitals, including Moscow.

How curious it then seems when visitors are struck by the weather. Is it as cold as they expected? Colder? The weather is fine, somewhat typical for these days, but the air and sea have changed. In 2014, ice would have formed along the fjord during the end of summer; now the water is clear of ice. The temperature is 5 degrees Fahrenheit, enough to make one's throat seize on every inhalation. The weather delights the visitors in the photo queue, as it grows to block the taxis and personal vehicles of residents inching their way to work or family. No one travels anymore; they arrive and depart. Travel is merely an act devoid of anticipation and experience. What happens in between is lost to eyes stuck behind a smartphone screen. Nowhere is this divestment from place more disappointing, as it is this very inattention that has left the arctic overburdened. Then the groups board their buses, still discussing the weather. Climate change is bringing out cloudberries and bringing in more people. It is also bringing about political instability for governments and their people alike.

"It's so good, it has to get worse," the bus driver says as the engine trills alive.

POLAR WAR

SINCE IT IS DIFFICULT to agree on what constitutes the arctic geographically, perhaps it is possible to agree that there are three arctics. There is, first, the arctic of someone who was born here, who takes nothing for granted and accepts its stark rhythm of life or death, of quiet and disquiet as inevitable, surmountable turbulence. Then there is the arctic of the person born elsewhere who arrives, and settles, as part of a personal and quixotic quest. Last, there is the arctic of the visitor, locusts devouring the region and leaving it worse than they found it. These are the bus riders—these visitors, here to glimpse the arctic of fables and history books, of postcards and glossy marketing brochures. Such images are easily had, taken, bought, exported. The visitors are here to make their own reproductions, rarely to experience them. One need only board a plane or two, the journey expeditious and comfortable, arriving on any number of solitary airstrips, to then brace oneself against the weather, and bring toward one's brow a camera or phone. Capturing a keepsake of the various regional tropes—a thermometer's negative digits; huskies tied to sledges; signs warning of polar bears and the dangers out yonder; the small aircraft moonlighting as mail carriers or ambulances—has never been easier. Then there are the exotic meats—puffin, Greenlandic shark, seal, whale, reindeer, musk ox—which may be seen in person hours before carving into them at a restaurant (sometimes with a Michelin star or two) that will also serve your favorite beverage chilled with real glacier ice.

Buses and road trips are a staple of arctic life. In Juneau, there is a paved road that abruptly ends after winding for forty-five miles toward nowhere. Residents called the drive they'd take there "out the road," as the road embodies both destination and journey, a poetic turnabout. In Keflavik, Iceland, the roughly hour-long ride to downtown Reykjavík nearly always begins with the driver standing at the front of the bus to address the passengers, asking everyone if they're "ready for Iceland's wonders!" Buses arrive and depart every twenty minutes, like clockwork. The bus in Longyearbyen is making stops at every accommodation in town. (I am left on one trip at the doorstep of Mary-Ann's Polarrigg—old miners' quarters converted into a hostel—and on another trip, off the bus route, abandoned to brave my own way the remaining hundred or so feet to a private apartment.)

As the visitors are deposited at their hotels across Longyearbyen, there is more chatter about the cold. About the need to purchase duty-free liquor and beer before the town's one store closes. About the incredible, disorienting half-light of near-winter. They are visiting an arctic of their own imagination, the arctic of an advertisement, embraced only as a hazy dream.

From one bus stop it's a short hike up a small ridge to the offices of Governor Lars Fause of Svalbard, who I find impatiently seated in a cushioned swivel chair at a large conference table overlooking the gin-clear Isfjorden. He wishes to paint a picture of circumpolar cooperation, of good times and clean living. However, a raft of activities falling just short of the threshold of armed conflict—known as gray-zone activities and part of the reason I am here, why perhaps the expeditionary forces are, too—undermine the credibility of his rhetoric. In no particular order, over the years since Gorbachev's zone of peace collapsed:

RUSSIAN SPECIAL FORCES USED SPITSBERGEN AS A BASE BEFORE PARACHUTING ONTO THE NORTH POLE, IGNORING THE DEMILITARIZED CLAUSE IN THE TREATY GOVERNING THE ARCHIPELAGO.

A MEMBER OF THE DEFENSE COMMITTEE OF THE RUSSIAN STATE DUMA SUGGESTED INVADING THE ISLAND.

CHINESE SCIENTIFIC INSTITUTIONS, KNOWN FOR THEIR MULTIPURPOSE DEPLOYMENTS AS INTELLIGENCE OPERATIONS, HAVE BEGUN EXPLORING INVESTMENT OPPORTUNITIES IN NEWER FACILITIES IN BARENTSBURG, A RUSSIAN-OWNED SETTLEMENT ON SVALBARD.

RUSSIAN DRONE OPERATORS FLYING IN RESTRICTED AIRSPACE WERE ARRESTED BY NORWEGIAN AUTHORITIES; SOME WERE

CHARGED AS SPIES, SOME WERE PAID HANDSOMELY BY THE STATE AS COMPENSATION FOR UNWARRANTED ARRESTS.

NON-SCANDINAVIANS IN LONGYEARBYEN HAD THEIR VOTING AND DRIVING PRIVILEGES REVOKED.

TURKEY AND CHINA USED THE NORTHERNMOST SATELLITE RELAY, STATIONED OUTSIDE LONGYEARBYEN, AND THE WORLD'S NORTHERNMOST RESEARCH COMMUNITY, NY-ÅLESUND, FOR MILITARY RESEARCH DESPITE NORWAY'S PROHIBITION ON SUCH ACTIVITIES, TOUCHING OFF A NASTY DEBATE ABOUT DUAL-USE THREATS.

THE RUSSIAN-OWNED SETTLEMENTS HAVE HOSTED RALLIES, FLOTILLAS, AND PARADED AROUND A REPURPOSED, WINTERIZED NORWEGIAN MILITARY VEHICLE IN SOLIDARITY WITH RUSSIA'S WAR IN UKRAINE.

A SUBMARINE CABLE CONNECTING THE ISLANDS TO MAINLAND EUROPE WAS SUSPICIOUSLY DAMAGED. TWICE.

"But of course, the cables are vulnerable. It's very easy to think that they could be cut off onshore or offshore," Fause deflects. In fact, the waters off the mainland conceal an array of sophisticated submerged cables used to detect illegal fishing and dumping and, among other monitoring activities, the presence of submarines. To imagine a vulnerability is knowingly unprotected seems a grave mistake. Fause has his suspicions, but, he says, with an absence of irony typical of Norwegians, the case is "frozen." He wonders if, having satisfied my inquiry, I will consider going for a hike or taking a tour, perhaps a beer-flight tasting at the world's northernmost brewery (more glacier water). Also on offer are sled-dog rides, though there is not enough snow—something that has become increasingly regular as Longyearbyen receives more annual rainfall with warmer temperatures. The dogs are instead lashed to low-riding wheeled carts.

Climate issues and tourists seem to mask a growing power struggle. It might have once been possible to disassociate geopolitical intrigue from everyday life here, but that was before Norway felt its sovereignty and the treaty were being exploited for geopolitical gain. This was a phenomenon that was itself a feedback loop. As the planet warms, more people wish to see the last of the icebergs (since, granted, the closest thing to an iceberg most will see in their lifetime is the lettuce), and to get there they take transportation that melts those icebergs faster. And as those icebergs and sea ice melt faster, nations increasingly explore once-frozen opportunities presenting themselves in the north, which brings more activity and threatens the fragile economic and national security of many nations, which melts more ice, and so on and so forth.

After the meeting with Fause, I walk over a footbridge connected to a utility pipeline and head for the main thoroughfare. There are few roads one might safely walk without a gun in Longyearbyen, so hazardous are the polar bears. This is a safe road. A clever visitor might think that so long as there is light, so long as night has not yet fallen—or that when it does they stay under lamplight—they can keep their eyes darting about them in order to stay safe. They would make every excuse to—rightly—calm themselves, remembering that all they had to do was reach a doorway for safety, assuming it was unlocked. (All doors in Longyearbyen only pull open; bears have not yet mastered door handles.) Rare, however, is the polar bear attack inside Longyearbyen.

At any rate, it is not impossible to find yourself in something bordering on a low-fi hallucination like the one I experienced with meeting Timofey's daughter: What was that sound over there—a gurgle of water or a growling predator? Is it afternoon or morning? Was I here earlier today or was that yesterday? How long has this sunrise lasted? Similarly, there are only two cafés at which one might sit down for a warm beverage. To limit the hallucinations, it is best to rotate between the two for the sole purpose of stimulating change.

At Huskies, one of those two cafés, Elizabeth Bourne sips hot tea. Timofey's daughter comes and goes, separated by the exchange of cloudberries and brief glances of recognition between the two women, and we continue our conversation. Elizabeth had moved to Svalbard from Seattle nearly seven years earlier, drawn by the light and community. Here, Elizabeth has been able to focus on her painting and artwork, making Longyearbyen her own and becoming the director of the Spitsbergen Artist Center. Elizabeth says the archipelago was a special bolt-hole—one into which she once could disappear, losing herself in a like-minded community at these extreme latitudes in search of a reprieve from the rest of the world. "If I'm going to experience the reality, I have to live here," she says of her first brushings with the arctic.

But this was no longer the arctic to which she had originally arrived. Geopolitics had made her life here more difficult. On this evening she is on her way to a meeting to discuss how life has changed for her and a group of residents calling themselves the "Unwanted Foreigners." They have been somewhat exiled, politically and socially, because of a desire, followed swiftly by actions and crackdowns, by the Norwegian government to prevent a foreign takeover of the island of Spitsbergen and of greater Svalbard. Part of the group's complaint is that they lost their right to vote in local elections. Officially, it was simply an interpretation of a long-standing law that had gone unenforced. Unofficially, Russia and China have made Spitsbergen a fixture of proposals for arctic expansion across two Soviet-era ghost towns. The Russian Federation was chartering its own boats from mainland Russia to Spitsbergen, on which Moscow owns and operates those two mining settlements beyond Norwegian oversight. In the spring of 2023, Russia announced that its mining company on Barentsburg, Trust Arktikugol, would build a research station with help from China, Brazil, India, and South Africa. China and Russia also signed a cooperation agreement between their coast guards to "combat terrorism, illegal migration, fighting and smuggling drugs and weapons, as well as [stop] illegal fishing," the head of the Russian Federal Security Service (FSB) Border Service, Vladimir Kulishov, said at the time.

In 2023, the government in Oslo formally restricted the civil rights of the "unwanted" non-Norwegian residents on Svalbard. Members of this group, who made up roughly 35 percent of Longyearbyen's population that year, overnight were made ineligible to vote in local elections, while residents from certain countries were also no longer allowed to drive their cars or trucks, their driving privileges suspended. There was little people like Elizabeth could do, except look elsewhere for a new oasis. She watched as non-Scandinavians like herself lost the use of their foreign driving licenses, making commutes and daily errands more difficult. She felt the reversal of the voting law was more reflective of a government in Oslo responding to geopolitical headwinds than to the realities of the civil coexistence between many nationalities on Svalbard. Meanwhile, she watched as local authorities ignored cases of wage disparities, workplace harassment, and spousal abuse among non-Scandinavians. The broader diminishing of democratic values was at odds with the Svalbard to which Elizabeth had arrived. For its part, Oslo refused to acknowledge the challenges to Svalbard's sovereignty and its growing security competitions there.

As a result of the foreign interference, Norway views all activities from outsiders with greater suspicion. They are existential threats to Norway's and Svalbard's security and prosperity. These fears are not Norway's alone. Across the circumpolar north many nations are increasing their military spending, expanding their fisheries and exploring oil and gas in their areas for fear of losing out on valuable resources. "It's kind of dualistic, but today we have a more heavy accent on everything in Russia and of course China. [The region's] always been very important to the Russians," Øyvind Nordsletten, the former Norwegian ambassador to Russia tells me. He recalls his time as a junior officer working in Moscow in the 1960s and flying out to Siberia to go fishing, a simple expedition. The plane landed not on a gravel runway but, rather, on a four-kilometer paved airfield. Why? It was for use by Russian bombers. Militarization "is a reflection on the fact that everyone's turning to the arctic as the new frontier in every respect—energy development and security policy and resources."

Despite that turn this was still the arctic familiar to many, whether in imagination or in actuality. Elizabeth uprooted her life to avoid the rest of the world, not to become entangled in its raucous indecision and bluster, its worries and anxieties, and its decisions exsanguinating from fear. There will always be the light, that half-shade of crimson that first drew Elizabeth. She has at least her memories of it and of expeditions to Greenland and Iceland, of circumnavigating the Svalbard archipelago. She nevertheless plans to return to America permanently. "It's time. I can see how things are going for non-Norwegians and it's going to get less and less pleasant."

Wars and infighting elsewhere were hitting people like Elizabeth and other "Unwanted Foreigners." They were cut off, apart. The world then seemed to halve. On the one side, there were those who sought a new world order rooted in fascist-adjacent rule: China and Russia, and their meddling in the high north. On the other side was a democracy-led standard and the status quo: America and its allies struggling to redefine what it means to be arctic nations in the era of globalization. Though climate was the ultimate cause for political domineering, the road ahead for some nations was still unnavigable. Many, the United States among them, had no road map, no defined plan for the arctic. They were heading out the road only to turn back, while it seemed for all the world that China and Russia and their allies knew exactly what they wanted: all of it.

FOUR DAYS AT
THE ARCTIC CIRCUS

THE WIND IS A thief trafficking in breath. These gusts, so strong and powerful and a potent reminder of gravity's role in keeping us grounded, are not accompanied by trash or debris. It is difficult to hear or see the bursts coming. Then they hit, bowling down from the Hallgrímskirkja—the Church of Iceland, bathed in shawls of amaranth light—sweeping past the hostels and hotels, into a residential area and beyond to the public bathing pool where children yelp and holler. Mothers glide along pushing strollers (almost everyone seems to glide in Iceland, as though skating on air), always battling a headwind. The RVs in a park adjacent to the pool shake and shimmer, but not from the gusts outside. Underfoot, cracks are forming and lava presses toward the surface.

It is early evening. The eruption alerts have trickled out steadily for weeks, hastening as autumn freezes into winter. The RV park erupts in its own sense, filling with the slow knocking of many diesel heaters attached to each van's undercarriage, ticking away time and fuel like metronomes. Across the open lot, a man sitting in the doorway of a converted passenger van—his temporary home—doffs his boots. He curls into the interior, clutching a bottle of wine, and slides closed the outside door with help from the wind and a fierce groundswell tremor.

In the morning, with no evidence of an eruption yet, the only disruptions are to visitors who learn the famous Blue Lagoon is closed as a precaution. A small town will soon be evacuated. Everything else goes along as scheduled: Ten minutes before the opening of a public hot-spring pool, a gaggle of men in various states of undress stand at the entrance and its sliding glass doors. Five minutes before opening, they inch closer, noses pressing against the glass, breath forming clouds like wet lichen against the doors. The young attendants inside—there are two—arrange a coffee station, ready some towels, and yawn at their computer screens. With sixty seconds until opening, the only people in a fervor are those outside, in the bleak cold darkness. The attendants are exacting, and they approach the doors to unlock them at the top of the hour, no sooner. Before the doors slide open more than a crack, the men squeeze through, pushing one another, never violently, but brushing so close to the attendant that it seems he is as movable and transparent as water.

Icelanders are a people deeply connected to the arctic waters that encircle them. They are fishermen who have pulled cod from the ocean to support the island nation for centuries, building it into a Silicon Valley of Cod, but it took a tragedy to teach them how to swim. Swimming is now a big part of life in Iceland, with one swimming pool for every 2,000 of its 335,000 citizens, thanks to the story of a steamboat named *Ólafur 57.*

Six children lost their fathers when the crew members, a mere 300 breast-strokes from the town of Akranes, a few nautical miles north of Reykjavík, and their captain had boarded a dinghy to make their way to church services. Then came the wave. *Hafið gefur, hafið tekur* (The ocean giveth, the ocean taketh).

That was in 1939. Within years, every community had a public swimming pool and children were taught not only how to swim, but also how to swim while carrying an unconscious swimmer. Nevertheless, death rates from drowning persisted largely because of the cold. In freezing water, the 1-10-1 myth holds that a person has one minute to control their breathing after the initial cold shock, then ten minutes to rescue themselves—get to a ship, an ice floe, something to float on. For those with facial hair, it is, as one wilderness survival instructor told me, advisable to freeze one's beard to that very object for what comes next: one hour before hypothermia blurs toward unconsciousness, then death.

The old men rushing past the attendant at the public pool are called something like *hurðarhúnn*, or "knobs," as in "doorknobs." They stick there, every morning. In a word, they are committed. They rush into the dressing rooms, place their clothes in lockers, and quickly scrub and shower before heading outside into temperatures well below freezing. They feel it only for a moment. Then they plunge into the lap pool, or the saltwater pool, or the various rings of pools heated between 60 and 100 degrees Fahrenheit by the natural springs feeding them from below. They enter, float, stroke a few laps, then shower and dress and are off to start their day, having perhaps learned that prevention (of ailments, of conflict, of deterioration) begins with action—rarely anything else. Iceland, with its sidewalks heated by those same hot springs, where nearly all the electricity comes from renewable energy, did not bend the land to its will. Instead, the nation bent itself to the will of the land. This is not only a recipe for nation-building in the arctic, but also one for climate stability and security in the arctic. The arctic is a land made not of people but of the environment, and those who attempt to impose their will upon it rarely succeed. *Hafið gefur, hafið tekur.*

THE SÆBJÖRG, THE MARITIME Safety and Survival Training Centre, is located aboard a ship next to the Harpa Concert Hall and Conference Centre in downtown Reykjavík. The training center works to improve the conditions for Icelandic seafarers, in part inspired by the ocean's impression left upon the nation's past. The conference center is in its own right a survival center, hosting an annual gathering tasked with managing, if not improving, life across the arctic. The attendees likely do not visit pools—though they may hit a lagoon or two—as they pass through town. They certainly will not learn what Iceland has learned: that stability in the arctic requires finding common ground, not only with other nations but also with the land itself. The annual gathering purports to understand these truths, but there is a historic distance, at no danger of closing anytime soon, between what people and institutions and governments intend to do in the arctic and what is actually accomplished.

On the night before the arctic's largest annual gathering of wonks, diplomats, delegations, and celebrities, the waves like daggers off the coast of the Reykjanes Peninsula, where the volcano still threatens to erupt, southeasterly gale-force winds cancel dozens of flights carrying participants. Others are delayed. Even the U.S. Coast Guard's all-star icebreaker is running late.

I finish a swim one morning and head to the conference center. I'm here to attend the 2023 Arctic Circle Assembly at the Harpa Centre, to which the U.S. government has sent its largest presence in years, with more than 160 personnel. The assembly can best be described as a cocktail hour for elevating the political clout of attendees who focus most on remembering not to charge their multiple rounds of drinks to the same bank card used for their per diem. Really, it is meant as a meeting of the minds, where serious people and serious discussions about a region in turmoil will take place. It is a different world from the one to which they say they cater. The entire four-day event is an exercise in contradiction. It is an apt, if painful, display of modern polar madness with a side of arctic hysteria. There are many national interests at the conference, all of which clash and rarely compromise. When a compromise is reached, like that which ended the Whisky War, it will usually be undone or disputed in the not-so-distant future.

In attendance are the Indigenous groups, like those in Canada, that want more federal funding. But the government in Ottawa, mired in scandals over reeducation programs and past abuses of Indigenous populations, is afraid of monetary solutions seeming like reparations and an admission of fault for years of institutionalizing native children. Ottawa is also more concerned with the majority of its population, which does not live in remote arctic territory. There are the Indigenous groups, like those in Greenland, that seek more autonomy from the colonizing government in Copenhagen. But Copenhagen relies on its colonial arctic territory in Greenland to remain relevant in global diplomacy and as an ace-in-the-hole for U.S. defense, which relies on Greenland for its northernmost military base. There is China, which claims to see the benefit of understanding climate change and wants a greater scientific presence in the arctic. But Scandinavian and American officials are leery that those aims are more about espionage and establishing a physical presence across the arctic from which China can expand its sphere of influence, as it has in Africa, potentially leaving northern nations saddled with unpayable junk loans. Though perhaps those efforts are less about financial manipulation than China's overall desire to establish a permanent political and economic presence in the arctic. Iceland itself is caught between wanting no part in geopolitics or the militarization of the north and having quietly allowed American military units to return after a prohibition of nearly twenty years.

The conference center is alive with representatives of myopic interests and bold intentions. I recall a high school graduation in Alaska I once attended where Tlingit performers danced and sang and crashed on their drums in mesmerizing unity. That seemed more exemplary of the region's legacy than this: servers toting trays of coffee, a live band playing modern soft jazz, and minifridges chock-full of Icelandic Skyr, catering only to outsiders. What is the purpose of the event? It is hard to say. Though created in response to the perceived insularity of the Arctic Council, the annual assembly does not on this occasion seem to bring the international community into the fold. It remains insular, almost unyielding in its abandonment of inclusivity. How is it possible to agree on something when that something is morphing, disappearing? Before one has a chance to determine what is changing, those changes are replaced with more.

Take the bewildering classification requirements for ice-faring ships as a measure of this challenge. One might choose to define a polar-class vessel by its brake horsepower or megawatt power. One could reliably choose the Finnish-Swedish or Canadian ice-classes scales. There is the scale provided by the insurance market Lloyd's Register. The American Bureau of Shipping (ABS) also has its own ice-class specifications. There seems to be agreement that ships intending to operate in the arctic should be categorized by the ABS's Polar Code classes 1 through 7, though this designation is rather new. The highest-rated ship on the Polar Code (PC) scale can travel unimpeded through year-round ice conditions, while those in the lowest class can safely traverse thin first-year ice, typically in the summer or early autumn. While the Polar Code's classifications apply to ships contracted for construction on or after July 1, 2007, they are often applied retroactively. Russia operates more than eight PC 1–2 icebreakers, thirty-one PC 3–4 icebreakers, and thirteen PC 5–6 icebreakers. China has at least one PC 3–4 icebreaker, and three PC 5–6 icebreakers, with many more in the design and construction phases. Canada and Finland have the bulk of other international icebreakers.

But what does it mean to say "first-year ice in autumn" when the thickness of that ice changes year to year? Convincing the public and politicians— specifically those in attendance at the assembly every October—that there is a need for icebreakers as the planet thaws is a tricky gambit. No one understands the complex inter-workings of arctic subject matter better than those here, yet, as with defining the arctic, even they cannot draw consensus on ship classifications. A theme of fluid indecision reveals itself.

The conference spans four days and its events are, as advertised, lavish; the conversations are, as expected, droll. Many attendants do not live in the arctic or thereabouts, but have much to say on the region and its global import. Rather than coming to listen, they feign empathy, make grand self-important statements, and reiterate the time-worn truth about the arctic's importance to a region whose people already recognize its importance, are not seeking solidarity from the outside world, and do not need outsiders telling them how to conduct themselves. This year's sponsor is the United Arab Emirates, whose representative recapitulates a tired stump into the assembly's echo chamber: "The melting of the ice in the arctic region must be stopped," to which everyone claps. Climate-change concerns offer carte blanche access for arctic stakeholder hopefuls, so long as it is feigned convincingly.

Over the weekend, the more than 2,000 attendees from seventy countries dressed in their suits and spectacular dresses sip cocktails while peering alternately at the Esja mountain peak or the puffins flitting about Engey, the island a stone's throw from the site of the *Ólafur 57* wreckage. Delegates retire to four-star Curio Collection hotels (there are two) or the EDITION (emphasis their own), where they meet in secret to discuss secret things about public peoples, places, and policies. At any rate, one might be so lucky as to catch a glimpse of Alaska Senator Lisa Murkowski's aides, trailing her in a hilarious malformation and glued to their phones, chiding requests from Indigenous representatives on their way to a meeting elsewhere. Welcome to the Arctic Circus.

THE ARCTIC CIRCLE ASSEMBLY hosts many nations because the future of the arctic has security implications for all. It was not lost on the Japanese when they landed on the Aleutian Islands during the Second World War, a time during which journalists were virtually denied access to Alaska and any news coming out was first reviewed by American military censors. Seven years before the Soviet Union tested its first hydrogen bomb in its own arctic region, President Eisenhower green-lit the creation of an early-warning system along the fringes of the North American Arctic, followed by the founding of the U.S.–Canada North American Defense Command. A year later, in 1958, the USS *Nautilus*, the world's first operational nuclear submarine, glided beneath the geographic North Pole. The United States saw its future national security potential in Greenland, which welcomed the USS *Nautilus* after its journey and briefly housed an underground American research station powered by a portable nuclear reactor. Norway, Sweden, and Finland also have long understood the security needs of their northern territories, which today bear the scars of Nazi occupation: In Norway, cement fortifications and bunkers stipple the northern fringes where the Third Reich once maintained its Atlantic Wall. Understanding the north's importance to a nation's security meant needing a forum in which security-adjacent topics could be addressed—a roundabout way of keeping talk of military security at bay, perhaps with the aim of rendering such talk unnecessary.

Shortly before the Berlin Wall fell in 1989 and the Soviet Union collapsed, a series of events after Gorbachev's speech in Murmansk unified the Global North and solidified, if only for a time, a collective approach to peace in the north. The NATO members adopted Norway's "high north, low tension" axiom for the circumpolar region. The Arctic Environmental Protection Strategy, signed above the Arctic Circle in Rovaniemi, Finland, in 1991, brought Canada, Denmark, Finland, Iceland, Norway, Sweden, the Soviet Union, and the United States under agreement over search-and-rescue operations in the Arctic Ocean, cemented cooperation on monitoring and assessment of climate change's impact on the north, and committed those nations to conservation efforts across the arctic. No fewer than three additional multinational conservation efforts were signed that year. Climate change replaced nuclear war as the driving source of concern for the world. As Gorbachev rightly noted in his Murmansk speech, the region was a "weather kitchen, the point where cyclones and anticyclones are born to influence the climate in Europe, the United States of America, and Canada, and even in South Asia and Africa." A 10,000-foot ice core plucked from the Greenland ice sheet revealed this truth through climatological data tracing back more than 100,000 years. Soon, studies would suggest that the Arctic Ocean could be ice free by 2040, the precipice of ice and snow atop the planet fully melted. But while climate was the reason for joint cooperation in the north, it would likewise become the antecedent to competition over resources and give rise to ever greater antagonism.

Five years after the Rovaniemi strategy, the Arctic Council was born. The forum, with the eight arctic nations as permanent members, further strengthened cooperation in the high latitudes. Many of the council's permanent members ratified the United Nations Convention of the Law of the Sea. The law defines the rights—commercial, environmental—afforded to nations within and along the world's oceans. The United States did not sign, believing the law would impinge on national sovereignty and limit future use of the Bering Sea and Arctic Ocean along which Alaska sits. The council was nevertheless productive, and in the intervening years saw Indigenous representatives and non-Arctic states—Italy, France, Japan, among others— join as observers. It also became the de facto home to representative bodies like the Arctic Coast Guard Forum. Resources found in the high north and cooperation therein were inextricably linked. Norway's sovereign wealth fund, an improvement on Alaska's own permanent fund model, grew to more than $359 billion by 2008. Alaska saw oil revenue of $8.5 billion that same year. The region, then at peace, seemed set to prosper.

Gauging what prosperity for all looked like was its own challenge. All the agreements and intentions and desires were and still are laid down in writing in dense policy papers, expressed in a jargon that excludes everyone impacted by these policies and positions. Does America want to be a good steward of Alaska and the Bering Sea, or does it want to exploit them? Does Sweden want to industrialize its northern territory at the expense of the Indigenous Sámi, or does it want to maintain what few cultural and heritage sites remain while increasing access to jobs and education? Is Russia militarizing the north for search-and-rescue missions, or are those bases the foundations for something more insidious? No one at the assembly really knows. No one can really, plainly, tell me, despite their assured confidence when discussing their proposals and the future of the region at this year's gathering.

SINCE RUSSIA'S FLAG-PLANTING STUNT in 2007, the conditions necessary for even a semblance of arctic unity have steadily deteriorated; it's a hollow collective marking the end of what many call "arctic exceptionalism."

Russia began to modernize its Cold War–era military installations across the arctic and increased its testing of hypersonic munitions there amid an uptick of military and commercial activity on the same northern islands where the Soviet Union had conducted its nuclear weapons tests. By 2017, China had sought funding for infrastructure projects across Scandinavia and Greenland, oftentimes blocked at the eleventh hour by Western intervention. India declared itself a near-arctic state in 2021 with the draft of its Arctic Policy, though it had been an Arctic Council observer since 2013. In 2022, Turkey began the process to join the treaty granting its citizens commercial and recreational access to Svalbard.

Not long before Turkey explored its northward reach, the war between Russia and Ukraine intensified, and a suspect weather balloon from China crossed into American airspace over Alaska. A U.S. Coast Guard patrol unexpectedly encountered Chinese and Russian warships seventy-five miles north of the Aleutian Islands. Where the United States had two working so-called heavy icebreakers, Russia had more than fifty; China had four in service and at least two soon-to-be-completed icebreakers to open up what it called its Polar Silk Road.

Compared to the more established Arctic nations, including Norway and other Scandinavian members of NATO, American efforts to address the increase in operations by unfriendly nations remains somewhat reactive and modest. Defense spending across Scandinavia has nearly doubled and in some cases tripled between 2022 and 2024. Militaries that once relied on conscription, a staple of the "total defense" homeland security strategies of Norway, Sweden, and Finland, have seen historic fluctuations in volunteer applications. Denmark, which partially rules over Greenland and the Faroe Islands, was recently forced to reevaluate its naval capabilities when it was discovered that the Royal Navy's battleships did not possess functional fire and control systems used to operate heavy weaponry like cannons and missile defense, effectively rendering their destroyers useless and vulnerable.

Meanwhile, the Distant Early Warning (DEW) Line and its series of radar stations across the North American Arctic—meant to detect enemy bombers and intercontinental ballistic missiles approaching over the North Pole—languished and quickly became outdated. When asked where a modernization project of the DEW Line's replacement—the North Warning System (NWS)—stood a year after the project was anticipated to begin, a representative from Canada's Department of National Defence tells me "it's too soon" to know, despite the availability of 38.6 billion Canadian dollars over twenty years, beginning in 2022. Infrastructure to aid climatologists in predicting typhoons off the coast of Alaska remains uninstalled, while homes and businesses fall to 100-year storms or rising sea levels, displacing entire communities. A Department of Defense

Inspector General report on the condition of American military installations across the arctic and subarctic—including Alaska and Greenland—
revealed substantial issues with the structural integrity of runways and
barracks. As China and Russia and so-called near-arctic states advanced,
the United States and Canada stumbled.

Since 2022, Washington has realized this and pivoted, formulating a
cohesive Arctic Strategy, revising and revisiting its substance across governmental departments as recently as May 2024. But its sclerotic response,
coupled with budgetary constraints and delays in key programs like the
Polar Security Cutter (PSC) program—which would endow the U.S. Coast
Guard with a fleet of much-needed ice-capable vessels—reflects a broader
unpreparedness and underinvestment in the region that can only hinder
American ambitions of reasserting itself as a major power in the arctic. The
PSC program alone faces nearly a decade of delays because of project mismanagement and a lack of funds. As one former diplomat who works on
arctic security issues tells me, "A strategy without budget is hallucination."

Policymakers talk about reinstating the arctic competence that was lost
since the Cold War. But development projects in the arctic are costly. Moreover, there is a long history of preventing Indigenous communities from
participating in those projects—so the government has shirked its broader
responsibilities to the well-being and security of its people, favoring con-
cession over cooperation. Few in Washington can readily see the value of a
region they believe is but a wasteland and home to only icebergs and polar
bears, oil fields and cruise ships.

In the last few years, the United States tried to redefine its own arctic region
and claw back some of what it has lost to atrophy: It slowly worked toward the
confirmation of the appointment of the first U.S. Ambassador-at-Large to the
Arctic; the U.S. Marines and Navy participated in the largest NATO exercises
since the 1980s in, of all places, the arctic regions of Norway and Sweden (the
U.S. Army went to Finland); the Department of Defense hosted an arctic dia-
logue ahead of an anticipated release of the revised Arctic Strategy; and the
State Department signed a flurry of Defense Cooperation Agreements with

Nordic allies in a sign that the region had become more hostile—though perhaps also as a hedge against the uncertainty that has accompanied the second term of President Trump, who denies the existence of climate change and threatens to withdraw the country from NATO. Previous administrations had similarly lacked consensus on their own arctic agendas. Nevertheless, as is characteristic of the Trump administration, it has a particular dearth of coherence about what, if anything, they seek to accomplish in the region, with even Senator Murkowski toeing her party's line when at best it serves to unsettle her constituents. Ultimately, the nation has a long way to go toward preparing for a conflict that began many years before.

"It is not hard to imagine that the arctic can be more than scientific research, especially if the Northern Sea Route opens," Admiral Rob Bauer, NATO's top military official, tells me one afternoon in a conference room at the Harpa Centre. "It is a concern. It is not entirely clear what the intentions are. In the military we need to be paranoid." As frozen ground thaws, a Cold War turns hot.

THE ARCTIC CIRCLE ASSEMBLY represents an effort to re-establish a spirit of cooperation despite rising tensions, if not a disconnect from the deteriorating relations across global northern territories. Founded in 2013 by then-President of Iceland Ólafur Ragnar Grímsson, former Prime Minister of Greenland Kuupik Kleist, and publisher Alice Rogoff of the *Alaska Dispatch*, it seems another initiative in a long line of well-intentioned spotlights on the arctic that has lost all relevancy. Over the weekend in Iceland, many talking points cite the need for cooperation and solidarity between those who occupy the circumpolar north. The arctic is, after all, shared. The first session is a brainy, drawn-out explainer on "Managing Risks and Relationships with Russia in the Arctic," during which the keynote speaker, Rebecca Pincus, the director of the Wilson Center's Polar Institute in Washington, D.C., outlines Russia's importance to the region, and talks about global cooperation. Only Russia, ostracized by the West even in scientific collaboration, is not welcome. Russia attends Arctic Council meetings only virtually, and only under select conditions favorable to them.

"You can still talk to [Russian] friends, but you can't be more than that," Terry Callaghan, the founder and once-chair of the Northern Forum, an Arctic Council observer member comprising mostly Russian and American interests, tells me one late afternoon near the close of the assembly. Booths and presentation boards are being put away as the event comes to an unceremonious close in the gloomy light of an arctic afternoon. "The limits are not on the Russia side, they're on the Western side. They need people on the ground. They have a lot to offer and to gain and we can't ignore half the arctic."

Much has been said about unity and agreement, yet there seems little of either in practice. One session's keynote panel consists of Indigenous representatives. The lecture hall is virtually empty. At that session, Sara Olsvig, the international chair of the Inuit Circumpolar Council, says, "We talk about tipping points" in the climate, but also within democratic initiatives. People nod. Some take notes. Then assembly-goers gather their Arctic Circle Assembly tote bags and meander to the next session, entering the maw of carnival gazers beneath the shards of the hall's elaborate geometric glass facade, lost in the cacophony of prophetic poetry masquerading as policy. Sometimes the story of the arctic is a mess of competing narratives.

Sometimes the story of the arctic is one of reminiscence, of how a boating accident led to the empowerment of a small nation, of how cooperation and unity could still be found if nations were to look into the past, to the days before lupine grew wild in Iceland, before someone brought the weed in a suitcase from America. *Lupinus nootkatensis*—known in its native Alaska and British Columbia as the "Nootka lupine" and as "lúpína" in Icelandic—was meant to revegetate Iceland, but instead some will say it diminished the natural splendors of the nation's volcanic deserts. The intentions were good. The outcome changed the face of a nation, and an invasive species reigns still.

Sometimes the story is a ballad to the present: to what cooperation has spawned, to the myriad voices now taking part in democratic circumpolar initiatives, to hope and inclusion, to how the lupines in Iceland have done wonders, turning dark fields into wavy pastures of purple like the beams of light cast against the Hallgrímskirkja. It is a matter of perspective. How a nation chooses to see its own arctic interests reflects how it feels about the global arctic's future. Cynicism is easier than dismay, but the conference was too much to bear. "Lots of talk," one colleague quips to me every October when the conference is held.

The inside of the camper van provides a greater sense of connection to this place that still taps into the imagination or reverence of the masses: the campground's proximity to the *hurðarhúnn*—who were actually doing something, clinging to their swimming pool doors, who are in some ways a living legacy of arctic symbiosis—offers a more holistic perspective on what the arctic is and still can be. It is a place worth defending to better learn about ourselves, to preserve our individual and national legacies, to place on display humankind's absolute best. Still, I recognize the chaos and disconnections and instability everywhere—the soft rocking of the eruptions underfoot, a nightly reminder that there are some things humans simply cannot control. The volcano will erupt in the coming weeks, displacing some 30,000 Icelanders from homes to which they might never return.

As I had so often during my time in Alaska, I retreat to a bar. In the north, bars can be both comfortable haunts and places of sheer haunting. On this particular night, a woman asks the bartender, Jon, who over several long stints in Iceland has become something of a spiritual North Star for me: "Where can I catch the northern lights?" Awards should be given for the alacrity of bartenders in absurd situations. He tells her all she must do is look skyward on a clear night. And after she's left, he gives more exact instructions: She should take a short drive down the road. When she reaches a parking lot, she'll find a nonexistent man named Frank who, for the price of 1,300 Icelandic króna ($10), will turn the lights on for her. If only desires could be had through will alone, through expectations, as though attainment of a goal comes through proximity.

"COMMUNISM IS
OUR GOAL"

NO OFFICIAL INVITATION CAME from Russia. Notes to the Ministry for the Development of the Russian Far East and Arctic (Minvostokrazvitiya) and the country's Far East and Arctic Development Corporation went unanswered. An unofficial invitation did, however, come from the Nenet reindeer herders on the Yamal Peninsula, in the Russian Arctic. The guide assured me on three separate occasions that the trip was safe, though I had my doubts because infrastructure integral to national security is located on the Yamal: major state-controlled oil and gas fields sit along the peninsula, the Russian government's Polar Express undersea cable project hugs the shoreline, there is a naval station for Russian nuclear forces nearby, and the region is home to many modern-day gulag-style prison camps. As we finalize details of the trip, the tour guide notes he will connect me with someone in the region, as he himself had fled since Russia's war began, fearing for his safety. On second thought, I graciously decline.

Before Russia's war in Ukraine, journalists and writers had on extraordinary occasions visited the Russian Far East, bringing notepads, pens, and a healthy skepticism. So, too, had a few researchers and academics. But the area is so sparsely populated and seldom visited by anyone else that it is often excised from maps (or, perhaps owing to potential, modern-day secret cities); even the government in Moscow began drumming up ways to entice its own people there, offering convicted felons—frontline cannon fodder in its war against Ukraine—homestead plots and cash if they made it home alive. If they returned in caskets, at least their next of kin might resettle in the northland.

Written accounts of the hard-won visits by the researchers and writers always take the same form: a brief rundown of the circuitous and maddening process of logistics and bureaucracy that granted them access to the highly protected and controlled region; an act of defiance against the rules and provisional access given to the visitor and then a rhetorical question about why nothing was done about the violation; a small insight into culture, geopolitics, or the inevitable pivot away from what the story was supposed to be about; an expression of leeriness to an untrustworthy subject giving an interview next to a picture of President Putin; a final ominous note on being followed, tailed, of having interviews and meetings canceled, and of casual (if amused) acceptance of threats by police or soldiers.

For all the literature produced by Westerners, most things about Russia are misinterpreted or misunderstood. Any Westerner will get it wrong simply by being from elsewhere (also true, as it is of anywhere, for anyone visiting the circumpolar north, writ large). Russia in the circumpolar north makes this particular topic doubly opaque to Westerners. What little I could glean from afar, and from contacts inside Russia, offered only more murkiness. The desire to understand should not be discounted, even when assured that understanding would never come. In a vain and supposedly hopeless attempt to understand Russia's arctic ambitions today, in light of the first major ground war in Europe since the Second World War and the first major military incursion into Russian territory since that same conflict, and as Russia dramatically increased its presence in the arctic, I called Nina Khrushcheva, the granddaughter of Nikita Khrushchev, the former premier of the Soviet Union.

Khrushcheva recalled a trip she took to Moscow in 2021. She attended the Victory Day celebration in Red Square, roughly a year before Russian tanks and troops rolled across the border into Ukraine. Amid the showing of green tanks and green intercontinental ballistic missiles and smiling female naval cadets in dark-green blouses, their hair pulled back in tight buns, there were for the first time white trucks, white tanks, depictions of polar bears. These were the northern battalions. What does Russia's ambitions in Ukraine have to do with the arctic? Why were Ukrainian drone attacks reaching Murmansk? Why were Russian bombing runs against Ukraine originating from above the Arctic Circle? What was the intention of Russia's arctic militarization, its promised development of the north? Was it just for show? These are the questions of a naive Western mind. "They are developing it. It's not to show off. . . . Everything is happening all at once. It's not from A to B to C. It's not a rational, Anglo-Saxon formula, it's not a cause and effect," Khrushcheva said. Russia's arctic ambitions, indeed its foreign and domestic policies, can be summarized in one word. "It's discombobulation."

Russia's approach to foreign and military policies, beginning with its annexation of Crimea in 2014, confounds and worries Western nations. Since then, it has not only refurbished and modernized dozens of Soviet-era military bases across its arctic territory, but also strengthened its fleet of advanced icebreakers. Russia worked with its partners in Africa, South America, and Asia to fund liquid natural gas projects and research centers in the north. Western countries have tried to blunt those efforts through sanctions and have developed similar projects of their own. The Western nations are adjusting their behavior based on the very haziness of Russia's desires—even more so since Russia has been placed on the international sidelines. Discombobulation is exactly the point. Keep your enemies guessing, in a state of constant darkness.

Russia's vision and strategy for the arctic is not international cooperation. When Russia was unable to achieve great-power status through its wealth of oil and gas, it tried to remake its playbook as one prefaced on fear and uncertainty. Exiled from much of the Western world, it has turned to Eastern partnerships to realize a goal it has had since Josef Stalin's "Red Arctic" propaganda, which throughout the 1930s sought to explore and greatly develop its northern territories. It is not so much a new vision as it is an ill-timed one, coming at a moment when there is little graciousness and goodwill for the government in Moscow and its strongman leader. That vision today seems aimed at securing a worldwide acknowledgment of Russia as the arctic's greatest stakeholder. That vision also seems to include a desire for the Northern Sea Route (NSR), which runs along Russia's northern borders and seas, to be recognized as a legitimate commercial channel. Artur Chilingarov, a former deputy chairman of Russia's rubber-stamp parliament and a Russian polar explorer, once put it plainly: "The arctic is ours and we should manifest our presence."

Western nations are allergic to discombobulation and declarations like those of Chilingarov. Given Alaska's proximity to Russia, any pronouncements of dominion in the arctic reverberate at home.

LONG BEFORE THE COLD War, Russia had a strong partnership with the United States in the arctic. It's easy to forget—despite one politician's enduring claim to easily see it from their backyard—that Alaska and Russia are at their nearest territories a meager three miles apart, the distance that separates two specks of trash-strewn and barely inhabited land in the Bering Strait. On one, Alaska Natives; on the other, Russian military personnel. The nations' mainlands are closer to fifty-five miles apart, across that same stretch of notoriously dangerous seascape, a fishbowl for some of the world's biggest fast-food and grocery-chain providers that is also home to Russian–Chinese naval and air force maneuvers. Between the two parcels, there are many differences, but also many similarities.

Alaska was once called Russian America. In the unsubtle diplomatic designs of William H. Seward, Abraham Lincoln's secretary of state, a treaty was signed shortly after the American Civil War, in 1867, securing a $7 million deal for the purchase of Russia's territory in North America. The modest valuation of the land, to an overstretched Russia still licking its wounds after disastrous military exploits in the Crimean War, reflected Russia's inefficient productivity across the territory. The Russian-American Company, Russia's first joint-stock charter, had been losing money in the Alaskan fur trade in the

1860s, with some 672 Russian settlers across Alaska's 663,268 square miles. U.S. officials saw beyond the paltry figures to envisage a more robust bread, wood, fish, fur, oil, gas, and meat basket. The company faced increasing competition from Canadian and American companies entering the sea-otter fur trade. From its loss, Russia also gained. The funds from the sale allowed Russia to bolster its Pacific defenses and further consolidate its empire.

American designs were similar—expansion, economic growth, domestic security—if not grander. Seward also wanted to expand, though not through armed conflict. He hoped to benefit from Britain's offloading of its overseas empire, extending America's arctic reach through Canada. "Peace is more propitious to the passion of empire, than WAR" and besides, "provinces are more cheaply *bought* than *conquered*," Seward said twenty-one years before the sale, which critics decried as the foolish purchase of an "icebox" and "polar bear garden." On the occasion of the sale, Senator Charles Sumner warned a weary Congress "it is difficult to see how we can refuse to complete the purchase without putting to hazard the friendly relations which happily subsist between the United States and Russia." There were fears that any transaction, no matter the potential benefit for America, could hamper relations with the czar, engendering a mutual distrust between the two nations for centuries to come.

After the purchase, Sumner proclaimed that "our Hyperborean Eldorado" may "seem to promise a new commerce to the country." That vision has been mostly realized: Alaska is home to abundant resources—oil, timber, minerals—and provides for the nation with only modest compromise. But Alaska is also home today to some of the highest national unemployment, and the highest cost of living per capita across all income brackets nationwide, on a state-by-state level. Nevertheless, one paramount vision of Alaska remains, if rarely discussed: The founding father of the U.S. Air Force, Billy Mitchell, would—seventy years after the purchase—declare that "in the future, whoever controls Alaska controls the world."

It was for this reason that the Russian empire, which was already a littoral arctic state and in the coming centuries would use its existing northern expanse (primarily Siberia) for imprisonment, industrialization, and exploration, felt the sale was nothing short of shameful. Indeed, Soviet historians later called the sale a geopolitical disaster. It would take less than a century for the losses resulting from the sale to be realized by Russia, and the gains to be likewise realized by the United States.

After oil was discovered north of the Brooks Range, in the North Slope, in 1968, a newly robust tax base offered Alaska and its inhabitants a fiscal lifeline, eventually leading to improved infrastructure and an economy dependent on tourism, fishing, and commercial timber. Nevertheless, food must be imported, as little of the state's potentially tillable land—3 million acres—is used for agriculture (though with temperatures rising, that acreage may grow). The majority of the state government's revenue comes from the oil and gas industry, followed by tourism, fishing, and federal transfers. Two squadrons of fifth-generation fighter jets, the largest concentration of advanced aircraft in the world, and dozens of military bases, radar stations, and observatories populate the state, bringing further federal funding and a rotating cast of military leaders and much defense spending. The lavishing dimes spent on defense rarely trickle down in significant ways. Villages are literally falling into the Arctic Ocean and Bering Sea, while delegations in Juneau and Washington, D.C., flirt with approving oil, gas, and mining leases in the state's largest wildlife refuge. Historically, these efforts benefit the few, rarely the many.

On the other side of the strait, in Russia, a confluence of the global shift away from carbon energy sources, diplomatic isolation, and economic insecurities has shrunken once-booming metropolises across the country's northern regions. In 1931, a group of gulag prisoners ventured to explore a vast coal basin in the Russian Arctic—terrain once inhabited by Indigenous Nenets. By the 1980s, the prisoners had established the city of Vorkuta, and its population soon ballooned to a quarter-million residents. Today, it hosts fewer than 60,000 and is shrinking faster than any other city in Russia.

About 2.4 million Russians live in the arctic, constituting more than half the global population above the Arctic Circle. But each year, about

18,000 residents leave the Russian Arctic, while three-quarters of the Russian defense budget (about $1.9 billion) goes to expanding the same region. (Since the collapse of the Soviet Union, Russia's Chukotka region has shrunk by 100,000 inhabitants.) Its social fabric in tatters, and its economy dependent largely on gold (once managed by a Canadian firm, which sold its Russian assets in 2022) and coal (by an Australian firm, which also withdrew from the Russian market), anything other than military infrastructure or natural gas exploration receives little in the way of help from Moscow. State energy companies like Lukoil, Rosneft, and Gazprom Neft seek to roll back environmental legislation limiting fossil-fuel exploration and production. Moscow embraces these requests, focusing on as yet nonexistent or wildly unfeasible technological solutions to climate change rather than decarbonization. A historical emphasis on energy infrastructure often cuts both climate change and communities out of the picture.

Most who leave the region today are replaced by foreign workers employed by Russian liquefied natural gas conglomerates and Russian military personnel, including a few marine mammals trained as spies. Those men who were recruited from Siberia to fight in Russia's war in Ukraine, and who have completed their service, are still being enticed back to the arctic by those state-funded land grants and promises of homesteads. Nina Khrushcheva said the Russian government is telling those veterans, "The arctic now is going to be the next potential theater of war and you guys have gone through Ukraine, why don't you become soldiers there?" The same could be said for Alaska, home to the largest population of combat veterans nationwide.

THE ONLY PRACTICAL WAY for me to glimpse the Russian Federation, then, was to return to Spitsbergen, where those two small settlements are still situated on historically sovereign Russian territory; the communities provide a modern portal through which to glimpse today's Russian Arctic. Barentsburg and Pyramiden are the center of both Russia's and its allies' attention to development (re: discombobulation) in the arctic. Nevertheless, the communities seem in the early stages of destruction. Russia would no doubt say—indeed, insist—these communities are undergoing a revival of sorts. Russia's vision of the arctic is one that has been remade over time, but how it manifests today is what largely exists in Barentsburg: a place of resource extraction, of political and strategic importance, from which to apply pressure elsewhere.

Barentsburg is 2,463 miles northwest of Russia's Chukotka Autonomous Okrug, or region. It is home to a Russian consulate, a brewery, and Trust Arktikugol, the state-owned mining and tourism company. For all intents and purposes, it is bona fide Russian soil, the red-white-and-blue tricolor waving over the town.

One day near the start of winter, the *Polargirl* powers east toward the settlement. The ship has been barred from taking visitors to the Russian settlement because of wartime sanctions, but it is going to Barentsburg anyway. The Norwegian authorities seem to look the other way, especially since Telenord has contracted the boat to bring telecommunications gear to the settlement. Rules are easily modified to suit one's purposes here. I decide to tag along. Aboard the *Polargirl* is a captain, captain's mate, and a smattering of others navigating the barren sea astride barren land beneath the shadow of the sun. A low quilting of clouds hangs overhead. The slurry, aerated wake of the boat fans out toward the Isfjord.

After little more than an hour, the *Polargirl* necks toward a dock. A heavy mist turns suddenly to needling snow. Ashore by the dock is a formidable, towering black mound of coal. There are no barges waiting to load it. Sanctions prevent the barges from docking here, too, severing whatever connections the community held with the southern world apart from the motherland. A small contingent of young men help tie off the boat, with her red-and-white hull, to the rust-chipped and snow-flecked truncheons. By the time the first passenger's feet touch earth, the snow is already resilient, resisting each footfall. A southerly gale washes the pass and the waves lick the shore, retreating briefly after gathering small tongue-lashings of snow. A wooden walkway runs parallel to the dock, away from the coal mound, up into the lower roadways of Barentsburg, a crescent of land in a pearly white desert.

Ascending the walkway, one is careful not to slip. The layers of snow rise past the soles into the bootlaces. The Inuit, Sámi, Iñupiaq, and Indigenous peoples of the north have dozens of words to describe snow, its different textures, and its rate of descent. Americans have no such language.

Lenin's face sits on a pedestal beyond the top of the stairs. Inscribed near him in foot-high Cyrillic lettering are the words "Our Goal Is Communism." The roads—there are two—are roughly paved and quiet. The yellow light atop a heavy-hauler flashes somewhere near the coal deposit. To the west stands a dark wooden chapel. To the east, a hotel and brewery whose windows are boarded up. In between are groupings of Brutalist block housing and a schoolhouse adorned with splendid murals depicting towering poplars in vibrant rays arching toward a celestial cosmos filled with stardust, the carbon offshoots of all the earth's humanity. More murals on the school building show fertile landscapes, children and women and men gallivanting about in uncomplicated, almost elementary existence. This mining settlement was once the idealized Russian society of the north. The murals seem to say, "Welcome to Hyperborea."

Only thing is, those depictions end where the paint dried. Barentsburg is a time capsule honoring hardship. It is both a symbol of the past, a reminder of what Russia once hoped would be its presence in the arctic—productivity, scientific pursuits, a political establishment away from prying Western eyes—and a reminder that those ambitions did not fall along with the Berlin Wall. That there remain stargazers believing the arctic still has something left unexplored, something left to conquer. Those ambitions here are still smoldering, waiting to be rekindled.

As we walk along the streets, deserted save for one or two young people shuffling to the hotel, I remember the words of China's leader Xi Jinping. During a meeting of the Communist Party a few days earlier, he had said, "Be mindful of dangers in the midst of peace. Get the house in good repair before rain comes, and prepare to undergo the major tests of high winds and waves, and even perilous, stormy seas." I steal a glance toward the wharf. Xi had recently declared China a near-arctic state, a newly coined term metabolized by every emerging nation worldwide. Russia followed those statements with its own grandiose pronouncements about the future of Barentsburg. Moscow pledged 1.5 billion rubles (about $23,986,500) to revitalizing Barentsburg and Pyramiden. It was the latest in a series of Russian investments toward arctic expansion that had started to worry Western nations. Soon Russia, along with its partners in BRICS—the informal organization of ten emerging world economies, many of which are hostile to the West—would begin talk of erecting a polar research station and headquarters here. The Norwegians have a phrase that might apply, one Russian resident of Pyramiden tells me: *Tung sjø*. She says it characterizes the troubled seas and currents below an otherwise calm and unassuming surface.

The Norwegian prime minister Jonas Gahr Støre reiterates the security situation in the north is unchanged. "We see no signs of an increased security threat in the North, not against Norway nor other countries," he says not weeks before our gaggle stands in Lenin's shadow. There is always a need to manage the expectations of a public desire for stability, and even more, for certainty that war would never reach that public. Certainly not here. Certainly not in a place where dauntless shadows move as if on a sundial, present only if exposed to light.

We bend forward against the diminishing autumn light and low gusts, crossing the settlement from its farthest east, the Russian Consulate, to the brewery, where hot and cold beverages await. Rubles are expected, other denominations accepted with a begrudging *spasibo*. Cash only. Taking refuge from the cold, I duck inside a small gift shop moonlighting as the post office. Trinkets hang from spinning racks and Matryoshka nesting dolls sit along a shelf in the far corner. Was that a nesting doll of Xi next to one of Putin next to one of Stalin? I earnestly look for one of Trump.

Posting a card to mainland Europe is a small act of charity: I am certain it will never arrive, and it never does. Afterward I linger a bit longer to marvel at the ephemera for sale, the emptiness of the place. Really, Barentsburg was all gravy. This town is mostly meant for exploiting the land; anything else it achieves, including tourism and a geopolitical thorn for the West, is a bonus. I worry little that the postcard or any other purchases will help fund the Kremlin's war in Ukraine. At the very least, it might help sustain one of the 300 souls still trapped in this frozen landscape, if only just for the cost of a lukewarm beer.

RUSSIA MAY HAVE ONCE thought to share the West's desire for cooperation in the arctic, but whether or not that is really what Russia wants is of little consequence. Its imperialist and economic vision for the arctic places Western nations into a defensive stance against it, whether that stance is a provocation in response to a provocation or merely a deterrence. What cannot be forgotten is the discombobulation between thoughts and actions, and the differences between Western and Eastern ideology. One Russian government official, purportedly reiterating the Kremlin's commitment to cooperation in the arctic, has said that "international law precedent does not define action." That by virtue of having committed to its role in the Arctic Council, by having adhered to the United Nations Convention on the Law of the Sea in arctic waters, Russia is a partner and a good-faith steward of the north. In the same breath, the official calls its wars in Georgia and Ukraine, and its meddling in Moldova and the Caucuses, "a separate discussion." This is the rationale of a strongman-led government, of perceived discombobulation. Their actions *there* do not reflect their intentions *here*. Call it selective jingoism.

That kind of thinking is alien to societies built upon democratic beliefs. In its 2020 Arctic Strategy, Russia feared its declining population; its inadequate social, economic, and infrastructure development; and a hobbled natural-resource exploration industry were threats to its national security. If viewed through a lens of discombobulation and diplomatic fog, perhaps the strong economic growth and international partnerships with India and

China and North Korea forged through increased reciprocity would help Russia realize its northern vision. However, Russia's invasion of Ukraine, its believed interventions in Western elections, and its attacks against critical infrastructure may mean that Eastern Europe is more of a threat than failing arctic infrastructure. It might also indicate that Russia thinks, as far as the arctic is concerned, it has already won.

"WE HAVE TIME"

ON GREENLAND AIR 628 from Upernavik to Qaanaaq, the island nation's northernmost village, passengers are now friendly. They first met on the flight from Kangerlussuaq to Ilulissat, then again on the flight from Ilulissat to Upernavik. They are not visitors to the arctic. They are coming home. The plane is a twin turboprop, its variable-pitch propellers transitioning on takeoff and touchdown for granular airspeed control. The engines whine and moan. At 20,000 feet, the cabin lights flicker on, then dim.

It is December 13, St. Lucia's Day. Scandinavian countries celebrate one of the earliest Christian martyrs by holding festivals and parades. Children dress in white and wear lighted wreaths atop their heads. It is something of a celebration ahead of the Winter Solstice, in the Norse tradition, a reminder of the nation's difficult history: Once it was home to the Vikings, then the Danish clergy, and then to the many years of colonization that followed. Before the year AD 1000, this holiday was celebrated with large bonfires to ward off spirits and to help realign the planet toward the sun once more. Today the pagan and Christian and many other religious traditions in December are still a celebration of that very thing: light's power to disrupt seemingly unshakable darkness.

Through the plane's window, there's a starry firmament overhead, with the polar winter one long night. In the southwest, a flash of sun. On a clear day or night, the American military base is visible on approach to Qaanaaq.

The base's lights illuminate nothing but a smattering of buildings and the airstrip, a reminder of arctic legacies and footprints, of how time is the great unrepentant equalizer. Supporting growth and opportunity only to abandon it has been the American calling card in the greater arctic for decades. America's derelict ambitions in the arctic are readily apparent across all of Greenland. At least since the mid-nineteenth century, the United States has been bullish on Greenland's potential to be an arctic surrogate for national defense. Then, between 2010 and 2020, those ambitions evaporated and with them went America's literal and figurative footing across the region, its ambitions overwhelmed by its inabilities against nature and foes alike.

Beginning in 1940, after the outbreak of the Second World War, American military planners, engineers, sailors, soldiers, and scientists arrived in Greenland. The first cars and ships and planes ever brought to Greenland arrived thanks to the Americans, who established twenty-one strategic outposts, including weather stations, refueling depots, and airstrips. By mid-1943, more than 15,000 Americans lived and worked on Greenland, and many would stay there long past the war's end. One such outpost at Thule, a lone weather-observation station manned by a detachment of four Danish soldiers from 1943 onward, remained more or less untouched, quiet, the province of a dogsled patrol with a mission to exert Denmark's sovereignty over its distant and long-neglected territory.

Qaanaaq, known as Thule throughout the mid-twentieth century, lies at the extreme northwest of Greenland. Its original inhabitants, the Inughuit, a subgroup of the already sparse Inuit, lived there for centuries, navigating the frozen expanse, a people of immense resilience. In the early 1950s, renewed American activity shattered the postwar calm. A new war was rising. The U.S. military, reimagining Thule's strategic potential as a cudgel against great-power competition, and only 900 miles from the North Pole, arrived during the height of the Cold War with the aim of establishing a new military base. In the decades that followed, scientists and the U.S. Army Corps of Engineers built a massive underground compound, called Camp Century. The cover for the project was scientific: They would drill ice cores to gather a deeper under-standing of the planet's geologic and climatic history, buried some four miles beneath the ice cap; these soil and ice core samples would raise material that had been hidden for more than 400,000 years. As a happy consequence, the military could also store a nuclear reactor there to help power the base, and maybe even house nuclear missiles in silos beneath the ice, in what became known as Project Iceworm.

That bounty of research and protection could not come without a price, America's safety and security always at the expense of others. Not only were there the physical, tangible land claims leased from Denmark and that circumvented Greenlandic approval, there was also the eco-logical issue of containing a nuclear dump site. Eventually, American interests in Greenland would eclipse the ambitions of Camp Century, spreading outward like the cracks that appear before a glacier calves into icebergs. The expulsion of the Inughuit to make way for the larger, more permanent Thule Air Base was swift and brutal. They were given only days to leave their ancestral homeland, forced to settle farther north in Qaanaaq in a place they had no choice but to build on their own, from the ground up. Even the cabin used by Knud Rasmussen, the legendary arctic explorer born in Greenland to half-Inuit, half-Danish parents, was relocated from Thule. It was 1943 when the planes first landed across Greenland, and they have not stopped since. Today, most travelers to Greenland touch down on and take off from airstrips built by the U.S. military, crosshatching paths in the skies above this arctic island shaped by the logistical needs of deterrence, posturing, and war.

A young girl no older than ten approaches the flight attendant. They speak quietly before the attendant closes the curtain dividing the cabin from the flight deck. I cannot say for how long they are gone, given the totality of darkness. Time is malleable, glacial, something registered only as part of one's interior. Soon the pair emerges wearing white T-shirts. In place of candles, they are holding cell phones, their flashlights pointed to the ceiling. The young girl sings in Greenlandic about St. Lucia as she walks slowly down the cramped single aisle. She rounds the back of the small cabin and returns to the front. She bows. The passengers applaud.

"We don't normally do it," the flight attendant tells me. "But we have time."

QAANAAQ, A COMMUNITY OF roughly 600, is barren when I arrive. The road from the airstrip is buckled and strewn with potholes, proof that time and growing concerns about the impacts of climate change are bearing out. Few lights flicker in the windows. I make my way to my lodgings. Past the snowbanks rising higher than the front door, inside there are books, nearly half of which consider American expeditions and exploits in Greenland, from Robert Peary's race to the North Pole to the enduring presence of American military assets. The displacement of the Inuit is a distant memory, but its legacy remains. The base's impacts are felt economically, culturally, and emotionally.

The base at Thule, renamed in 2023 as the Pituffik Space Base, is no longer home to thousands but, rather, a small contingent of the Danish Armed Forces (among the 200 Greenlandic and Danish contractors stationed there, there is the Sirius Patrol, the twelve-person dogsled patrol that travels the ice cap for months at a time) in support of the U.S. Space Force detachment there. America's arctic territory mainly lies in Alaska, which has more than 34,000 miles of coastline and houses dozens of military sites; Thule is its lone counterweight, geographically adjacent to America. A week mulling about Qaanaaq provides me an opportunity to reflect on the way time has altered some perceptions of the region—its utility and viability—while leaving unchanged some fundamental truths about commitments toward longevity.

Located 1,100 miles west of Spitsbergen, Thule is home to reindeer and arctic temperatures so cold that Fahrenheit and Celsius often meet as equals (at 40 degrees below zero). It is the sole security guarantor over what is known as the GIUK (Greenland, Iceland, United Kingdom) gap, a thorny approach to the Atlantic Coast, and the homes of 36 percent of the American population. The base, its contingents from the U.S. Space Force and visitors researching the arctic region, serves as a major space surveillance and satellite command logistics hub, and it plays a key defense role in providing early warnings of nuclear and over-the-horizon attacks. Only, like much else in the American Arctic, it is relegated to an afterthought. For now, the base, used for passive intercontinental ballistic missile (ICBM) monitoring, is virtually defenseless and relies on Canada for the ice breaking needed for resupply deliveries in winter. America's arctic issues, Mark Cancian (a senior adviser with the Center for Strategic and International Studies' International Security Program) tells me by phone as I call around to contacts near and far, are not a hindrance but, rather, an opportunity for the United States to spread some of the burden sharing to allied countries. "I think that the forces and resources are limited. We're not going to be able to rebuild the kinds of capabilities we had during the Cold War," Cancian says. "It's hard for the Europeans really to replace the U.S. The best they can do is complement us. But [the arctic is] one place where they are arguably ahead" and it's a place where the United States should follow, rather than lead.

"NATO capabilities are reasonably strong for surveillance both undersea and in the air," James Stavridis, a retired four-star U.S. Navy admiral who was the supreme allied commander at NATO from 2009 to 2013, tells me, "but not strong in terms of ice-breaking, which is of course crucial, or response to ground-force movements."

One of the reasons the base is defenseless is climate change. When the base was built as an extension of Camp Century, located some 215 miles southwest of the base's location today, it was primarily intended as a first alert against nuclear attacks. "We were laying flat slabs of concrete on top of permafrost, without really understanding the scientific implications or the engineering implications and challenges that that might present," Colonel Brian Capps, a former Pituffik base commander, told me in our first conversation. "We have had a lot of decades to learn and adjust and do a lot of infrastructure rebuilds and infrastructure tweaks," Capps said in 2022, shortly after assuming the post and inviting me to come for a tour. Meanwhile, the barracks that house Space Force service members and researchers has sparse or nonexistent wireless access. When asked about improvements to the runways and living conditions, Capps said, "We have started."

During their visits to arctic and subarctic bases in 2021, staff from the office of the inspector general of the Defense Department found those issues and more. At a base in Alaska, inspectors placed their handheld walkie-talkies into the cracks splintering across runways and splitting concrete foundations to show their depth and width—fissures that would sometimes swallow the devices whole. The fences at Joint Base Elmendorf-Richardson, in Anchorage, are riddled with gaping spaces along the bottom, thanks to shifting permafrost. And that permafrost, which is thawing due to increasing summertime temperatures, makes the perimeter penetrable to anyone willing to get down and crawl. At Pituffik, in July of that year, walls had been spreading apart at a rate of ten centimeters annually, while doorways were lifting off their thresholds. Rust, a phenomenon of wetter climates, was rampant. The runway was fracturing as a result of constant thawing and refreezing of the permafrost, coupled with pooling water on the runway after each thaw. A grim picture emerged of those bases and of troop preparedness that inherently impacted the military's broader domain awareness.

When I follow up on the invite to the base, Colonel Jason "Shrek" Terry, who took over command of the base and its 821st Space Base Group, rescinds the offer, instead choosing to answer my questions only in writing. "The base continuously assesses and maintains the facilities, which is a significant effort in such an extreme environment in the Arctic," Colonel Terry wrote. "Military construction funding is a multi-year process. If a structure is damaged due to settling, it can take time to secure funding to repair or replace damage." Colonel Terry was reassigned after one year and a new commander was appointed.

That's the thing about the arctic. America seems terrified to engage with those who are interested and also lacks the commitment to stay. Time is the only real commodity here. With quick deployments, the U.S. presence is always one of fickle commitments and evaporating knowledge, which is surprising for a region where you never meet anyone only once.

THE BASE AND ITS history, and its importance to the security of America, was a subtle if understated presence in the 2022 National Strategy for the Arctic Region, the first arctic strategy since the Obama administration, which focused mostly on Alaska and threats from Russia and China to the state and its territorial waters. The new strategy, and the more than half dozen reports, strategies, and doctrines that followed, prioritizes security and lists Russia and China as the nation's primary threats in the arctic, while emphasizing the importance of cooperation with arctic countries like Finland and Sweden. It was also vague and lacked a commitment to real action. The first subpoint under the security pillar was to "improve our understanding of the Arctic operating environment," placing the United States far behind its stated competition—countries that are already building more icebreakers and military bases, planning ice routes, and stationing troops in cold weather, not occasionally ferrying them to and from the region. The report says nothing about icebreakers or the importance of expanding U.S. reach within the greater arctic region. And while it did broadly address climate change and increased regional competition, it didn't say anything about how those issues would be dealt with. A lack of collaboration with Russia is an impediment to the researchers working on the cryozone or arctic projects—something the report failed to address. Furthermore, the report said an American presence in the arctic region will only be "as required," a designation that many arctic experts felt was shortsighted.

The reaction to the report from some arctic analysts I spoke to was underwhelming. Heather Conley, a Deputy Assistant Secretary of State during the George W. Bush administration and an expert on arctic affairs, told me the nation's new policy remained amnesiac and reminiscent of years past. She said it did not reflect the changing geopolitical and commercial importance of the region. "I see policy that in its isolation is fine," Conley said. "It's just fragmented and doesn't necessarily have an overarching policy objective that everyone understands. . . . And they're not reflecting these really important geostrategic economic and security shifts, and how are we adjusting policy." Conley still thinks the United States sees the arctic more as a domestic issue—an arena where policy should focus on natural-resource extraction in Alaska and offer whatever assistance is requested by Indigenous populations—than an international one.

"As an Arctic nation, the United States remains deeply committed to the region," a State Department spokesperson wrote me in an email in 2022. "As noted in the recently released National Strategy for the Arctic Region, the United States is advancing efforts to mitigate and build resilience to climate change and ecosystem degradation. The State Department looks forward to advancing this vision through a whole-of-government, evidence-based approach, including with science agencies and in close partnership with the State of Alaska and Alaska Native Tribes." To be fair, this is the way government was meant to work, best as I can tell: A department of government releases a report with stated objectives, not specifying how to achieve them, then experts in several departments throughout the executive and legislative branches create more documents with slightly more focused objectives, with the hope of receiving funding for intangible, though qualitatively adjacent goals. Then Congress meets. Funding may or may not be allocated. And when the money comes, it is rarely spent well, or it is simply not enough.

Meanwhile, America's arctic infrastructure and military bases languish. The center of any visible U.S. arctic strategy for now is Pituffik, America's northernmost and forgotten military installation.

Until 2024, the only deep-water port in the American Arctic—necessary for staging large military expeditions abroad or to transport troops quickly by sea—was at Pituffik. And like each of America's six arctic military bases—all "contingency bases," meant to be staging grounds for expeditions into the arctic—it may soon get a long-awaited makeover. Through NORAD, the United States in conjunction with Canada is updating its northern partner's largest air base to replace Cold War–era early-alert detection systems for nuclear missiles. Congress had already approved funding for a deep-water port in Nome, Alaska (which, like every American project in the arctic, would be seriously delayed), while the U.S. Navy has flirted with reopening its own former base in Adak, Alaska, and has sought to open a deep-water port in Utqiagvik, Alaska. While Nome's port is under construction, Pituffik could again host strategic-bomber operations and fighter-pilot squadrons, as well as static anti-aircraft and anti-ballistic missile defenses. It could also serve as a launch point for more surface ship missions above the Arctic Circle, which the United States has rarely done since the end of the Cold War. There remain plenty of exigent problems with available solutions but instead the U.S. government is playing fast and loose with new infrastructure, rather than getting its current assets in order. None of these considerations were mentioned in the White House's arctic plan (the National Strategy from 2022) explicitly, or in the other plans that followed.

INFRASTRUCTURE IS A CURIOUS thing. It is not just the footing and grounding of a family, or the promise of an uninterrupted education, or the ability to navigate the landscape and seek out sustenance unimpeded. Nor is it the assuredness that comes with a hospital that stands firm atop an active layer—a green mattress of spongy soil and sedge grass, which is the shifting, thawing, and refreezing layer above multi-year permafrost. It is that those buildings and outposts, almost without exception in the arctic and subarctic, are important for their dual-use capabilities. A hospital is also a community center. A fisherman's wharf is the disembarkation point for tourists. A satellite dish can monitor space or the earth and the people around it. A military base is never only about defense but also about deterrence.

One of the reasons the White House and the State Department, in its latest arctic strategy, called China a "pacing challenge" in the arctic was this: China's increased use of dual-purpose scientific-military platforms across the high north to establish a presence and to gain more information about the region. "Beijing is talking about opening up more globalized international governance and moving away from the Arctic 5 and the Arctic 8 dominating the region," Trym Eiterjord, a research associate at the Arctic Institute with a focus on China and Asia in the arctic, tells me, referring to the five Arctic Ocean littoral states and the eight Arctic Council members. For at least a decade, Beijing had compared the Arctic Council to the Monroe Doctrine, arguing for a multi-tiered approach to the arctic, a grievance the West has long ignored. "Whether it comes to liquid natural gas ventures or ship building, the main variable when it comes to China as a threat in the arctic is its cooperation with Russia in general."

Over the last decade, China has sought to gain a competitive advantage across the arctic. In Greenland specifically, its aspirations have been clear: It sought to fund infrastructure projects and mining operations in a nation that is said to hold large and rare quantities of oil, iron ore, and precious metals like rubies, cryolite, and uranium. China's own infrastructure across Africa has left nations in debt bondage; this is a concern in Greenland, too. But the workers who would have been brought from China to support mining operations in Greenland that are now on hold have also been a source of anxiety, especially in the arctic, perhaps even more than Trump's own entreaties into the region.

It is widely understood in the fields of security and intelligence that anything that can be used for gathering signals intelligence will be, whether or not the host or organization or ship captain knows it. Perhaps those people would simply rather not know anything. Plausible deniability not only allows for the dual-use of infrastructure and people, but also permits those responsible for it to ignore its endgame. Qaanaaq's airstrip and the IS18 satellite relay in town connect more than just geography; they bind together histories of displacement and defense and neglect. The IS18 stands as a reminder that time and space here are governed by forces far beyond the local community. For years, American interests have been secured with each flight, each missile, and each relay tower, their strategic importance outweighing the voices of the Indigenous peoples who remain. Denmark and Greenland, forming a unified front, sought a modicum of acknowledgment from the Americans about their commitments to those northern communities when Space Force took over the Pituffik base from the Air Force. That is why America renamed the base after the traditional Greenlandic name for the region. That was all it felt was necessary.

CHRISTMAS LIGHTS DRAPE THE exterior wood paneling of the brightly colored homes around Qaanaaq that appear like stepping stones toward the polar heavens. Trash bags near burn pits sit mere feet from a home's porch. Toilets in much of rural Greenland are simply seats over industrial-strength garbage bags. In every bathroom, there are industrial-strength twist ties to close the toilet bags when they are full. I walk for little more than an hour, reaching a small footbridge over a shallow runoff, and I turn left when in the darkness I see the shadow of a young boy dribbling a soccer ball just beyond his front porch. He looks a young teenager and practices different kicks. The ball is partially inflated. He passes it to me, the ball skipping over the ridges of last night's frozen snowfall.

Time has not been kind to Qaanaaq. It is indifferent to schedules and planning, and budgets, and human desires. Time, and the way it weathers arctic infrastructure, is both vexing and expensive. It is expensive because it takes time to debate whether mitigation efforts are more costly than adaptation. This is not a problem unique to the military. Yet military personnel and the U.S. Army Corps of Engineers have spent years trying to educate leaders in Washington and across the military about the threats posed by climate change. The same people who built Camp Century are now going unheard. It is likewise vexing because the infrastructure in question is not merely this building, that outpost, or the other thing. It is everything.

As we kick the soccer ball, the IS18 satellite relay looms large atop a nearby hill. It is a quiet sentinel dispatching signals and anchoring the invisible tethers that tie this remote place to the global infrastructure dedicated to American defense. The IS18 is more than just a relay; it's also part of the wider architecture of U.S. military influence—technology perched at the edge of the world, guarding against unseen threats from space and beyond. Its presence here is an extension of the same strategic logic that, decades ago, pushed the Inughuit out of their homes to make way for what became Pituffik. The satellite is meant to bring better internet connectivity to the region, but just as the arctic fox sheds its orange coat for white, nothing here is what it seems. The IS18 relay, though ostensibly for climate research, could be used for much more.

For the Inughuit, time and flying have always been linked—by the planes that pushed them away and the satellites that now cast long shadows over the places they call home. In nearly a year, the new airport in Nuuk will be completed and visitors to Greenland will be able to fly directly to the nation's capital, over distant radar stations and silent outposts—a project once nearly funded by the People's Republic of China, with all the political, security, and economic strings that come attached.

"But look, it is no secret the way China operates, right?" a senior State Department official told journalists in April 2020. "And they have tried in the past to—what's the word I'm looking for—wiggle their way into Greenland in unhelpful ways by acquiring critical infrastructure that would be problematic for the United States and our NATO allies and, of course, the Kingdom of Denmark. That includes efforts to purchase in 2016 Gronnedal, which is the old American naval base there constructed in 1942 to protect our flight routes during the Second World War. The Kingdom of Denmark shut down that deal. And more recently in 2018, an effort to get involved in the airport construction in Greenland in three locations."

It truly is no secret how China operates in the arctic: Communist Party members have said of Sweden that "for our enemies, we have shotguns"; of Norway, that it "must pay the price for their arrogance"; that Iceland is "ill" and "weak." Greenland was another way into the arctic, and mining, in the past two decades, and tourism, in the past five years, have become essential inroads for foreign investment from allies and enemies alike.

Denmark stepped in after the United States pressured it to foot the bill for the airport, offering a loan that outbid the Chinese government's, and blunting China's goal of becoming "a polar great power." In doing so, its dual-use efforts have increased as long-standing American infrastructure has weakened. China has in the last two decades deployed the use of military satellites for arctic-only purposes and has frequently deployed submarines and unmanned underwater vehicles in the polar regions. The ongoing American presence at Pituffik, for all its shortcomings, remains necessary.

Large monitors in airports across Greenland still celebrate in patriotic commercials the Royal Danish Navy and its boats, though incapable of protecting against anything, given their inoperable weapons-control systems. The everyday Greenlander wants economic stability and diversity, and to remove themselves from their reliance on Denmark—on any nation hindering their independence. Polling suggests that Greenlanders treasure direct foreign investment and might seek to rejoin the European Union, prompting Nuuk to strike a mining deal with a Danish-French group as Trump sought mineral rights, so long as that cooperation does not jeopardize their enduring commercial and security arrangements with the United States, by which they mean they are open for business but not for sale. Mostly, these things do not come up in daily discussions. What is spoken about is an ongoing and historical frustration with the Danish treatment of Greenlanders. And their desire to lead more comfortable lives, like they see being done in Europe and North America. With Denmark's thumb on export controls and on deciding what is to Greenland's defense and security and economic benefit, achieving this end may be difficult.

During my walk, the howling of the dogs in Qaanaaq comes on eerily and all at once. Their leashes are staked into the land along the shore. A headlamp bobs through the darkness along the icy rim. I nearly step on a dog, silently curled into itself and chained to a bolt-hole invisible through the snow. I walk the length of the harbor beneath the noontime stars. Beyond the edge, boats, trapped like icebergs, sit motionless while adjacent mountain peaks reach northward, beyond which the blueish veil of the southern sun pokes at passing clouds. No landscape is visible beyond the footsteps before me, the glow of my headlamp consumed by the darkness. I check to see if it is even lit. A man at the breakfast nook in his home cranes his head to watch me pass. I wave. He squares himself in the window. He waves back, presenting a toothless smile, then darts away and is gone. Later, just after midnight, boys with their arms around one another's shoulders, and trailed by four young women in a casual lockstep, make their way down the street toward the harbor.

They veer right, jolly together, time less something they engage with and more something that occurs naturally. How strange it then seems that those of the south gather in crowds but do not touch, that we must be taught to share, that the only way to know another is to gather intelligence from afar.

ON MY DEPARTURE FROM Qaanaaq, the flight attendant crochets, comfortable in an empty seat. Through the windows is that permanent sunrise on the horizon, a reminder that were one to stand at the geographic North Pole, one could, simply by walking in a circle, divine their own sunrises and sunsets. The flight attendant's name is Pernille Ploug Larsen. Her fingers are ringed with tattoos representing the Mother of the Sea.

Greenland Air once prohibited its staff from displaying tattoos, despite the ink being a significant part of Inuit culture in this country of roughly 56,000. Most people think she's an alcoholic just because she's from Greenland. For the same reason, they believe she is bitter toward Denmark. It's disappointing, this inherited outsider's view of Greenlanders. She's in favor of independence, which Greenland continues working toward, though it still relies on a massive yearly block grant from Denmark to support its economy. Pernille is twenty-six and a fan of romance novels. She buckles up for landing, and the pilot quips that some landings, like this one into Ilulissat, are like landing on an aircraft carrier: short runway, mountain or cliff dead ahead.

Despite the nation's size (three times that of Texas), most people live and work in a few select communities. When we bump into each other next, Pernille is no longer in uniform. She is snuggled into a handmade scarf at the airport in Kangerlussuaq, heading to her partner's place in Denmark. Then we board our direct flight to the only international destination possible from Nuuk: Denmark, the home ruler over an isolated people beholden to—though tussling with—the powers who came and conquered and stayed.

"A WARM WELCOME IN COLD WATERS"

THE *HERMES* MAINTAINS ITS course as its crew unfurls a pilot's ladder on the fishing vessel's starboard beam. Steel cables behind the 226-foot boat slowly reel in a mesh trawler, briny with sea water, back onto the deck. The trawl behind the *Hermes* is the first obstacle a small crew aboard a dinghy, launched minutes earlier from a nearby Norwegian Coast Guard vessel, are mindful to avoid. Waves thrice the size of the small boat rise and settle all around. Within the belly of each swell and the churning Arctic Ocean waters, both the *Hermes* and the KV *Svalbard*—Norway's preeminent icebreaker—disappear from view. Gulls surveil overhead. The small boat pulls alongside the *Hermes*. Then comes the jump. The first coast guard officer scrambles to the bow of the dinghy, called a sea bear, grasping the railings. Erik (who at nineteen years old is the KV *Svalbard*'s de facto small boat captain) motors the sea bear into position, maneuvering the bow perpendicular to the *Hermes*'s hull. The first officer counts the rise and fall of the bow. At the highest point, he lunges. Erik jams the throttle into reverse, clears the ship's wake, and circles back. The officer dangles from the pilot's ladder and begins a slow and deliberate ascent. The fishermen help the officer swing first one leg, then the other, over the railing as the vessel creaks and crashes against the sea. Three officers and myself board the *Hermes* in this way, jumping from the sea bear onto the rickety rope ladder amid ocean spray and minus 4 degrees Fahrenheit gales, with only the Arctic Ocean to catch us if we fall.

71

Like many of the Indigenous peoples of the north, Norwegians have vocabularies dedicated to the slightest variations in wind and sea conditions. The sea state on this day is near gale. Or, a *sterk kuling eller liten storm*. Or, *nording kuling*. Westerners have hardly such a language, are mostly passersby in the arctic, and in a region such as this might find themselves unable to speak. They would become mostly deaf.

The decks aboard the *Hermes* are oily. The comingled stench of salt water and fish permeates the air topside. Dried cod, called *torskjaeringer*, dangle from railings and banisters, over ladders and ports around the ship, stiffening in the wind. Inside, the Norwegian Coast Guard, or Kystvakt, officers doff their cold-weather survival suits and climb a rusty ladder to the galley. They are unarmed, save for a waterproof case filled with tools to measure the spacing of gill-net webbing and the crew's catch. The Kystvakt officers are looking for whether the crew is using the right size net for its intended catch. Are they releasing illegal fish trapped in the net? Does the crew have proper safety and life-preserving equipment on board? And perhaps most important of all, is the captain reporting accurately how many fish he catches? Overfishing is one of the greatest threats to arctic ecosystems. Though it is possible even these boardings, which send the Kystvakt down into the belly of the boat where the fish are frozen and stored, do little to protect from overfishing. And when this crew was deployed onto Russian fishing boats suspected of severing the submarine cable outside Svalbard, they could prove no wrongdoing.

The captain's wife sits at a table in the galley, calmly spreading strawberry jam on one layer of a sponge cake. Today is the captain's birthday. The officers accept coffee but not cake, then climb another ladder up to the bridge. The captain, a heavyset man in his early fifties, lounges in a large plush chair mounted to an adjustable stanchion and surrounded by more monitors than the KV *Svalbard* has on its entire bridge. Fish-finding radars, weather radars, a monitor with swirly, colorful lines depicting the tracks of other nearby vessels and their fishing pots, high-wide transducer outputs, and radios and satellite phones in various configurations and states of disrepair. The year before, fishing vessels had accounted for 44 percent of all unique vessels recorded in arctic waters.

Fish factory ships, like the *Hermes*, counted for 20 percent. Overall, the number of total vessels in arctic waters had nearly doubled from the decade before. In the American Arctic, that number is expected to increase by sixfold before 2030. Not unlike its arctic infrastructure, America has also lagged behind in its defensive and patrol capabilities, hence all the talk about more icebreakers. It has lost its arctic competency since those heydays in Greenland. Its qualifications are in shambles, though its partners, like Norway and other Scandinavian countries, prove to be exceedingly capable—touchstones from which Washington can learn. But they cannot continue doing the work of world policing, of monitoring illegal fishing and submarine cable attacks, on their own.

Two Norwegian officers place the waterproof case on a desk at the back of the bridge near the net drum operator's terminal, by a window overlooking the fence gallows. Lieutenant Andreas Soløy, who is never kind to his consonants, speaks with the captain through a perpetual nasal drawl. Lieutenant Soløy asks for the captain's logs. "Youp, youp," the captain responds and the routine inspection begins.

TO BECOME ACQUAINTED WITH these waters, I traversed them for two weeks aboard the KV *Svalbard*. The days were long and busy, as there is much activity across the ocean. The Arctic Ocean is home to at least 240 native species of fish (include the adjacent seas through which some of those fish migrate and that number rises to 633), some 29 species of birds and seabirds, and 12 species of marine mammals, as well as ongoing oil and natural gas exploration that is, almost twenty years later, still largely based on a study by the U.S. Geological Survey from 2008 that has not been revisited since its publication. Much of what is being sought after in the region—its waters and continental shelf—relies on this report, and it is this report that has itself caused a sort of race for natural resources at the expense of the existing biosphere.

The Arctic Ocean is otherwise covered half the year by a heavy layer of ice, up to fifteen feet thick in some areas. The ocean is slightly less than 1.5 times the size of all fifty states, consuming 5.4 million square miles, its waves lap-ping at 28,203 miles of the littoral coastlines of Canada, Greenland, Iceland, Norway, Russia, and the United States. It is the world's smallest and shallowest ocean, but it is also the least well known and is largely unmapped. Each year the ice cover changes dramatically—a mercurial blue-and-white map redrawn every summer.

How a vessel may navigate these waters is also in seasonal flux. Where that vessel travels is somewhat more finite, as there are three routes a maritime vessel might take in the arctic. There is the Northwest Passage, along the northern fringes of Canada, mired annually by ice that has gradually thickened over many winters in labyrinthine straits whose coastlines are dotted by massive permafrost-cored hills called pingos. Some summers, the straits are still frozen and unpassable. Canada is not interested in export traffic, so the passage is often undertaken only by expedition or passenger vessels, with some resupply ships reaching coastal Indigenous communities. There is the Transpolar Route, altogether impossible for any regular vessel to navigate, as it would require carving a pass through the North Pole and the ice that still gyrates there year-round—at least until 2030, when climatologists predict the ocean may be ice free in summertime. (It certainly will be by 2050.) And there is the Northeast Passage (sometimes known as the Northern Sea Route, which is encompassed by the larger Northeast Passage), along the northern shores of the Russian Federation. As the planet's warming makes this route more viable with each passing year, taking it would save shipping vessels some eleven days in their routes from Asia to Europe—offering an alternative to the Suez Canal and the dangers of the Gulf of Aden and the Red Sea. Today the Northeast Passage is the most expedient route between Shanghai and Rotterdam, and if further utilized, it could bring a major influx of traffic from Europe through the Bering Strait past Alaska. Unlike the Northwest Passage, which recorded its first commercial transit in 2013, the Northeast Passage saw 31.5 million tons of cargo—mostly liquid natural gas and crude oil—shipped along the route before the war in Ukraine. In 2022, despite sanctions and a tepid Western blockade of Russia, the route handled 34 million tons. Putin demands even more cargo traffic.

The Northeast Passage sees relatively few vessels that are not Russian or Chinese. Most ships traveling it are military, or they are ferrying supplies and personnel to floating oil rigs, or they are fishing or catering to tourists. (Berths aboard the Russian nuclear-powered icebreaker *50 Years of Victory* for a thirteen-day roundtrip from Murmansk to the North Pole start at $32,000.) Sanctions restrict most international vessels from operating in those waters, which require a tax paid to Moscow, which in return supplies its fleet of icebreakers to escort vessels through the passage into the Barents or Bering Seas. Navigating the ice and the weather of the north requires its own set of risky calculations, even without Russian interference. Essential are a crew trained in cold-weather survival and the ability to overcome superstructure icing, which occurs when the spray from the ocean creates large ballast-like ice chunks that weigh down the ship.

Even as the Arctic Ocean warms, there may still remain a need for ships with ice-breaking capabilities. Thin ice invites those who may not be equipped for the voyage. Recreational and commercial vessels may be willing to tempt seas they had not previously navigated, for example. Transit traffic in the arctic has been modest in part because there are few vessels with a suitable classification for traversing ice, but that may soon change. The only reason for ships to be there is for the fish, and even the fishing vessels, in the recurring cases of damaged submarine cables, are suspect.

After the inspection aboard the *Hermes*, which turns up a few small violations, we head back to the KV *Svalbard* against a diminishing sun and rising seas. "When the ice melts and gets thinner, then it's more necessary to have icebreakers because the activity will go farther north. There will be a situation where they can't come out of the ice, just like Greenland," Commander Geir-Martin Leinebø tells me in his office aboard the KV *Svalbard*. Only days before, a luxury cruise ship had become stranded above the Arctic Circle in waters off the coast of Northeast Greenland National Park, an area roughly the size of Spain and France combined. The Royal Danish Navy, which provides national defense to the semiautonomous nation of Greenland, took four days to reach the vessel; a fishing trawler and a research vessel worked together to tug the cruise ship free. Even nations well practiced in arctic operations find their capabilities limited simply by nature and distance, and the chasm of time. "You can see," Commander Leinebø says, "it will increase."

THE KV SVALBARD IS Norway's premier offshore patrol vessel. It is the nation's largest, and the first Norwegian vessel to reach the North Pole. She is beautiful. Outside, a sullen gray hull matches the arctic skies, and hard angles suggest the deflection of radar. A flight deck above the fantail and a cannon on the front are overshadowed by a superstructure (including the bridge) that spans the ship's width at 62 feet and extends nearly a third of the ship's 340-foot length. On outings from the boat there is not one crew member, from private to officer, who does not marvel at the ship, her stature and might, always stopping despite the swell to snap a selfie of themselves with the icebreaker as a backdrop. Inside, the KV *Svalbard* is spotless, its interior of white and polished metal railings and ladder treads clean and orderly, resembling something more like a cruise ship than a man-of-war. Not a length of internal plumbing or wiring is visible. Nothing is chipped or cracked or broken, but were something to chip or crack or break, one of the *menige* (pronounced meh-knee, meaning rank-and-file, or privates, many of whom are conscripts) would see immediately to its repair. The KV *Svalbard*'s chef worked at Vaaghals, a Michelin Guide–rated establishment in the trendy Barcode neighborhood of Oslo. There is a softwood-paneled sauna. The crew dotingly calls her "Hotel Svalbard."

"Have you see the Lider class?" Commander Leinebø gushes one day. On his computer, he clicks through renderings of Russia's newest fleet of icebreakers, fawning over the 120 megawatt power of each new ship. The KV *Svalbard* has just 5 megawatts. America's one icebreaker serving the arctic, the U.S. Coast Guard Cutter *Healy*, a 22.4-megawatt ship, was laid down (construction begun) five years before the KV *Svalbard*, in 1996. The

"A WARM WELCOME IN COLD WATERS"

U.S. Coast Guard had plans for several new arctic and subarctic patrol vessels, but the program, much like Norway's own, remained hampered by design flaws, inflated costs, and political headwinds. In short, the excitement and hope of new ice-breaking vessels has thawed.

The *Healy*, however, possesses a few toys Commander Leinebø wished the KV *Svalbard* had: multi-beam sonar, an A-frame davit, and unlimited Starlink wireless internet connectivity. What the *Healy* does not have—aside from the cannon and KV *Svalbard*'s unique mandate to patrol joint waters where European Union, Norwegian, Russian, and Faroese fishing fleets catch cod and shrimp almost year round under an agreement that falls outside of wartime sanctions—are the sharp ice-breaking propellers, mounted on 360 degree pivots, which allow the KV *Svalbard* to sheer ice chunks in reverse. "It's a very good thing," Commander Leinebø says. Even with their fancy propellers, ice navigation still proves complex. The cost of real-time ice images from space, at some $2,700 per image, is prohibitive. So the crew relies on open-source imagery, a cost-saving approach to assist in the navigation, allowing the KV *Svalbard* to plan its route in advance and navigate cracks through both fast ice and pack ice.

Though the bulk of KV *Svalbard*'s mission on this patrol is to provide search and rescue in the event of an emergency, to project Norwegian sovereignty through its presence, and to uphold maritime laws through fishery inspections and unannounced boardings, the crew has a secondary objective. It will, in a week's time, rendezvous with and escort the *Healy* as it completes a voyage across the top of Russia. The voyage, billed as a scientific mission, will underscore the difficulties of arctic navigation and the challenges presented by an increasingly hostile north—the secondary and more concerning product of a warming Arctic Ocean. It will reveal the tensions in the arctic's strategic competition, and also American skittishness about working in higher latitudes in today's climate roiling beyond a new ice curtain. The geopolitics and weather did not bother the Norwegians so much. They were prepared, joyful even. "We are the welcoming committee," said

KV *Svalbard* operations officer Lieutenant Captain Kim Aasland. "A safe escort into Tromsø, a warm welcome in cold waters."

NORWAY IS AMERICA'S GREATEST partner and asset in the arctic. To spend time in Norway is to see not only Europe's arctic ambitions but also America's and Russia's. Norway is NATO's listening post in the arctic, the country having increasingly become something of a linchpin in Washington's greater arctic ambitions. The U.S. Marines deploy to Norway for four-month exercises in cold-weather warfare training each year. A U.S.-funded radar system snoops on Russia's nuclear military assets on the nearby Kola Peninsula. The U.S.-made RQ-4 Global Hawk surveillance drone will soon be stationed at a new long-range base on the northern island of Andøya. An early-warning system against cruise missiles in the High North—the name given to Norway's arctic territories—it is the first of its kind outside North America, and was funded in part by American taxpayers. The significance of the rendezvous of the two icebreakers was a notable enrichment of that relationship. What was not clear was whether the significance of the meeting point, at 82 degrees, 35 minutes north, was recognized by the crews. This was nearly the exact location reached during William Edward Parry's fifth expedition some two centuries before, the farthest north anyone had then traveled.

For all their cooperation, the United States relies more heavily on Norway than Norway relies on it. And this being their first interaction, the Norwegians will learn firsthand that, for all the American might and posturing elsewhere, they are ill-prepared for the realities of a changing arctic. They can learn much from Norway, their language about the ocean and its north winds. They will choose not to, though, despite shared goals and adversaries.

Paramount to mission success in the north is cooperation. The Norwegians have maintained for decades their "high north, low tension" approach to working with the Russians to their east, alongside whom they cooperate on fisheries management and search-and-rescue operations at sea. Norway created a military buffer zone, barring foreign troops from basing their units too close to the country's border with Russia since 1949. Each year, even during the war in Ukraine, Kystvakt officers and Russian Coast Guard officers routinely embed on each other's vessels. "You see, the Russians, they're trying to open the Northeast Passage, and they will manage it, because they have a lot of icebreakers. But what kind of icebreaker have we? KV *Svalbard* is twenty years old. What has the USA? Nothing. But there is another side, another message," Commander Leinebø says. "Putin is saying, 'I'm the boss in the arctic,' and he is."

A FEW NIGHTS AFTER the inspection aboard the *Hermes*, the moon continues its waxing. Every fifteen minutes, the crew on watch hits the deck to do ten push-ups ("160 push-ups a shift!" one officer exclaims); between sets, they talk about their hometowns, josh over local dialects, aimlessly tie and retie knots into ropes, and fumble with a Rubik's cube. Tins of smokeless tobacco are accompanied by endless mugs of coffee. In the days ahead, before reaching the sea ice, the crew enlivens the mostly silent arctic waters. There are the guns (40mm cannon, 12.7mm machine guns, 9mm pistols), the thrum of four Bergen diesel generators, the rancorous exertions associated with the physical fitness of fifty men and women playing volleyball or lifting weights, the outboard jet-propulsion motors of the smaller sea bears used to conduct fishery boardings and inspections, the crunch and crackle of king crabs prepared on the fantail. Oscar, a lifesaving practice dummy, appears at random places around the boat each day, frightening unsuspecting passersby. Over breakfast each morning, Commander Leinebø addresses the crew through an intercom system, known as a "pipe" in maritime jargon, to challenge the complement of sailors to brain exercises. At the end of each patrol, whoever answered the most quizzes correctly wins a trophy. Everyone is relaxed. Lieutenant Ocean, an officer who recognizes the suitability of his surname to his chosen profession, compares it to "going camping at a cabin for six weeks with friends." The crew is comfortable in the isolation and the north. This is their home. Then come the sounds of grinding ice against the hull.

The KV *Svalbard* and its crew cross the eightieth parallel, deep within the domain of Boreas Rex, King of the North Wind, shortly after a dinner of cod and potatoes and spring rolls with duck sauce, with few glancing out the portholes to see the wavy chop and lingering bergs outside. Above and beyond are smooth and shapeless clouds drawing their curtain. The horizon blurs, a silken velum of blue-white titanium where sea welcomes sky. The sky offers an effective, if archaic, way of navigating icy waters. A savvy captain can decipher what lies ahead simply by inspecting the clouds: blue means open water, white means more snow and ice. In the evening, two large floodlights illuminate the water ahead and the KV *Svalbard* edges farther north. Ice slides up onto other sheets across the ice pack, creaking and splitting, slipping and sliding, like a deck of cards being shuffled. Howling and thundering claps wail against the hull. Everything trembles and rumbles, explosions like Howitzer salvos. While steadily making our way to the rendezvous with the *Healy*, there is no peace anywhere.

Where the KV *Svalbard* and the USCGC *Healy* plan to meet holds national and historical significance beyond Parry's 1827 expedition. The Norwegian explorer and savant Fridtjof Nansen once purposely stuck his boat, the *Fram* (literally, "Forward"), in the polar sea ice in 1894 and flowed with the ice vortex for two years. He is a Norwegian legend, a neuroscientist (nerve endings still bear his name) who eventually became the League of Nations' first High Commissioner for Refugees. There was a passport issued in his name, the Nansen passport, which assisted some 450,000 stateless people after the First World War, and for which he won the Nobel Peace Prize. Like most seafaring Norwegians (admittedly a tautology, for nearly everyone I meet while staying in Norway for two months was studying marine biology or fishery management, joining the navy or coast guard—sometimes by conscription, sometimes by choice), the KV *Svalbard* crew revered Nansen. The north was in their DNA.

Though the ships and crews had never met, the KV *Svalbard* had once assisted the *Healy* in 2020, after a fire destroyed two propulsion motors and their shafts, waylaying the boat near its home port of Seattle. Its mission had been to collect scientific moorings in the Beaufort Sea, north of Canada. The National Science Foundation (NSF), which supports polar science missions aboard the *Healy*, instead called the Kystvakt to collect the instruments before their batteries died and with them two years of irreplaceable data about the changing north. A seven-month race-against-time through the Northwest Passage ensued, during which time the ice pack could have frozen early, trapping the KV *Svalbard* in ice it could not break. The crew prepared for this eventuality and steamed fast. They arrived on schedule, with time to spare. The KV *Svalbard* retrieved the moorings and returned home.

After a week at sea, in the darkened wheelhouse of the bridge, with the two large spotlights leading the way through the ice outside, the *Healy* radios to say it is having issues and will not be at the meeting point (*metapunkt*, in Norwegian) on time. It will not be the first or last time the *Healy* calls to delay. No matter the ship, breaking four feet of ice is slow business.

Later that night, still on the KV *Svalbard* bridge, the crew is joking and giddy, and breaking several feet of ice at a steady clip. Then the *Healy* calls again.

"Standby for MRSAT call," a voice crackles across the bridge.

"Standing by," a *Svalbard* officer responds.

The bridge falls silent. Officers exchange befuddled glances. One of the junior officers shrugs. Darkness outside.

Someone breaks the silence and says, "What's a MRSAT?"

The formality of exchanges with the U.S. Coast Guard shatters whatever levity the KV *Svalbard* crew held until now, like being caught in admiration of something private. The Norwegian crew is casual at sea, comfortable in its domain, its backyard, speaking with other boats without an air of ceremony. Such formalities felt out of place. The KV *Svalbard* was where it promised to be. The *Healy* was not. For all their ceremoniousness, the complement of sailors aboard the *Healy*, by all metrics America's vanguard in arctic waters, will soon dispel any notion of American exceptionalism or professional capacity in the High North.

"IT'S SO UGLY," ONE of the *menige* tells me as the *Healy* pulls behind the KV *Svalbard* to follow it out of the sea ice. The private is right. The *Healy* is all right angles and is often referred to as "the largest and most technologically advanced icebreaker in the United States." This may well be true, but the bar for those commendations is objectively low. It is one of four, and the only American military vessel routinely in the Arctic Ocean. It looks like someone had the hull of an icebreaker in storage and decided, on a whim or as a prank, to park a small housing development atop it. The superstructure containing the bridge, the science deck, the cabins and quarters, the galley and hospital bay, and the onboard gift shop (stocked with *Healy* memorabilia and cigarettes, among other sundries) is massive—a white elongated eyesore poked through with portholes and seated upon a red hull adorned with a variety of cranes and rigging, and that A-frame davit Commander Leinebø so admires. The *Healy*'s displacement, at 16,000 tons, is nearly three times that of the KV *Svalbard*, and it is eighty feet longer and twenty feet wider. Much like the KV *Svalbard*, two rigid-hulled inflatable boats (RHIBs) hug either side of the *Healy*. Despite its greater size, the *Healy* radiates an air of inferiority when the KV *Svalbard* sails alongside it. A man-of-war she is not.

Then again, the *Healy* was not designed for defense or for law-enforcement operations. It was built specifically to support science missions, one of the U.S. Coast Guard's congressional mandates to reinforce climate research. The *Healy*'s crew is at the end of its five-week mission, having departed from Alaska in late August to travel west through the Bering Strait into arctic waters. Throughout their mission, the crew recovered and redeployed moorings installed in 2021, like those the KV *Svalbard* helped collect in 2020, as part of the Nansen and Amundsen Basins Observational System, which gives insight into how water from the Atlantic Ocean is introduced into the arctic at the shelf water level, deep-basin interior, and upper ocean; and to help develop an understanding of water circulation in the region. The crew also managed forty-one conductivity, temperature, and depth casts, sampling the seawater to investigate its properties at various depths in the water column. In short, they were checking up on the thermohaline circulation, among other projects concerning that great ocean conveyor belt.

ONCE OUT OF THE sea ice, the sea bear and one of the American small boats begin transferring crews between the ships. Some of the KV *Svalbard* crew has been with the *Healy* since it left port in Kodiak, Alaska, more than forty days earlier. Now the *Healy* will board some of the *menige* to give them a tour of the ship and bring aboard officers, among them Lieutenant Ocean, to lodge with the crew for the rest of their trip through Norwegian waters. But the marine off-load release—the device connecting the American small boat to a crane and winch—is stuck, frozen stiff, caked in a thin layer of ice.

"Looks like you need a mallet," I say.

"I *am* a mallet," comes the petty officer's response as he hacks away at the device with a piece of flotsam. The American crew seems ravaged, tired and frustrated. A U.S. Coast Guard spokesperson offers an explanation: "The arctic in particular is a harsh environment that is apt to weathering equipment." Be that as it may, the truth is that the *Healy* was not built for small-boat storage and deployment in that climate. The Norwegian vessel has two wells on either side that cradle the sea bears, keeping them ice free and ready at a moment's notice. The petty officer's comment is nevertheless representative of the American approach to the arctic, its incorrect valuation of its competency, and a woeful impracticality where a lack of planning precedes poor results. A month in trying conditions would weather the best of crews and equipment, but there were still many things—like maintaining a deicing slurry—that might be prepared onboard or in advance.

During the transfer between ships, I recall a cartoon I saw in the engine control room aboard the KV *Svalbard* explaining the Dunning-Kruger Effect, a cognitive bias leading to the overestimation of one's abilities. A blindness to one's true competency, beliefs, and behaviors can stymie growth and effectiveness. While the photos and maps mounted along the passageways in the KV *Svalbard* are secured to the walls at four corner points, the frames mounted in the passageways aboard the *Healy* are hung on the walls only by a single mount, the images and accolades listing in time with the ship. The beige interior of the *Healy*, where exposed piping and wires sway from ceilings and bulkheads, is more reminiscent of a Second World War battleship revived for parade duty than of a vessel primarily charged with scientific missions, so it is a stark contrast to the KV *Svalbard*'s unblemished interiors.

Trouble has plagued their voyage. The *Healy*'s onboard gyroscope and X band radar consistently display errors and false readings. An outbreak of the coronavirus swept through the *Healy* early in its voyage. A flood rose through the chamber where the crew stored the CTD array (named for those conductivity, temperature, and depth tests, though known onboard as a "rosette"), which is arguably the leading reason for the mission and for which contingencies have not been made. Not long after this voyage, another fire damaged the *Healy* on its annual arctic patrol in 2024, preventing scientists from conducting research "operations in more than 7/10s first-year ice," as though Boreas Rex himself was cursing America's northward endeavors.

"It's been basically OK," says Lieutenant Commander Gudmund Johansen, one of the KV *Svalbard* officers who had joined the *Healy* in Kodiak, as we stand on the *Healy*'s bridge. He is silent for a moment. "For the most part."

The *Healy* has been to the North Pole on several trips, a complement of privately and publicly funded scientists aboard each time. That was until recently, when the U.S. Coast Guard announced at a meeting of the Arctic Icebreaker Coordination Committee that general science missions would now take place every two years, rather than annually, to accommodate more research missions of a different kind: those undertaken by the Naval Research Laboratory and private military contractors working under Pentagon contracts.

Onboard the *Healy* this voyage are members of the U.S. Army, U.S. Navy,

U.S. Air Force, the British Royal Navy, two Kystvakt officers, members from the Carderock Division of the Naval Surface Warfare Center, and more than a dozen scientists on government contracts. There are researchers studying the circadian sleep schedules of those working in the High North (a team from the Naval Postgraduate School distributes sleep tracker rings that monitor REM cycles); ice hull designs and the operational capacity of non–ice ships in ice (Memorial University of Newfoundland, St. John's); global navigation satellite systems and technology (Mitre Corporation researchers, confidential); and high-visibility colors for safety equipment. The bridge and weather decks are a cornucopia of server racks, antennae, and computers.

Their mission this August and September, however scientific in nature, cannot avail itself of politics and posturing. Commanding Officer Captain Michele Schallip took command of the *Healy* only a few short months before embarking on the trip, a freedom-of-navigation tour and scientific expedition similar to those undertaken during the Cold War. Along their route they have been trailed by the Russian icebreaker *Akademik Fedorov* and a Tupolev TU-95 bomber, as the *Healy* traversed the far northern stretches of the Chukchi, Siberian, and Laptev Seas into the Arctic Ocean, before sailing alongside the KV *Svalbard* between Russia's Victoria Island and Norway's Kvitøya. In that area, as the two boats sail toward Tromsø, Russia issues a notice-to-airmen warning of military activity. Russia asserts control over Franz Josef Land, an archipelago that is Russian territory to the east of Spitsbergen that overlaps with Norway's exclusive economic zone. One of the Naval Postgraduate School researchers tells me, "There is always a hidden war around a country's EEZ."

From the very start of the tour given to the KV *Svalbard* crew, the welcome is less than warm. When the *menige* gather on the bridge, Captain Michele Schallip turns to an aide and asks, "Do they speak English?" This seems an indicator of how vast the gulf between these NATO partners has grown. An engineer scoffs when the ship's 1,220,915-gallon fuel capacity does not elicit excitement from the conscripts, and he tactlessly fumbles when he tries to convert standard to metric (4,621,000 liters). Lieutenant John Thompson, the medical officer aboard the *Healy*, quips after giving the conscripts a tour of the medical bay that he "could have told them it was a flux capacitor and they would have just nodded," referencing a fictional technology from *Back to the Future*. Packets of jelly beans and Haribos and Marlboros litter the floor of one berth. The chefs prepare a dinner of corn dogs.

A HANDFUL OF SCIENTISTS and sailors remain in their berths on the morning of September 29, 2023, most stricken by seasickness as the *Healy* emerges from the ice and follows the KV *Svalbard* toward mainland Norway. Desk drawers aboard the *Healy* slide open and slam shut. Chairs relocate themselves from one end of a room to the other. Anything not bolted down, and there is plenty that is loose, hits the floor with crashes, thuds, clatters, and clanks. The *Healy* was built for ice navigation, with a rounded hull that offers poor stability in open seas. Outside there is nothing but sky and a fleet of fishing vessels cruising through a dull light dimmed by a thick curtain of fog. For a while there is silence, the soft padding of waves against the hull.

The silence rises to a murmur as a vessel that has been tracking alongside the KV *Svalbard* and the *Healy* throughout the night has still not appeared on the radar. Captain Schallip has no concerns, she tells me later. Sometimes vessels have issues with their electronic beacons while at sea. Then Captain Schallip receives a call from the bridge. She goes to the helm as the fog recedes. A lookout spots an aircraft over the starboard bow. No plane, friendly or foreign, has been expected or has been seen in the area recently. As the fog clears, the vessel that is tracking alongside the two ships is identified. It is a vessel from the Russian Coast Guard of the Border Service of the FSB, and it is closing within a few nautical miles of the ships.

An announcement comes over the pipe. "VIPER drill, VIPER drill." The environment aboard the *Healy*, tense before, stiffens. The VIPER drill, or Visual Information Personnel, is part of an all-hands task for military entities to properly document unsafe and unprofessional interactions with foreign militaries or vessels. Several sailors grab cameras and record the interactions.

The Russian ship comes within one nautical mile and launches a helicopter. Though the KV *Svalbard* has one aircraft hangar and the *Healy* has two, both are vacant. Budget cuts and resource reallocations have stripped both ships of this fundamental tool for search-and-rescue missions, a mandate of coast guards the world over.

Mason Schettig, a team member from Ship-based Technical Support in the Arctic, a project funded by the NSF's Office of Polar Programs, steps onto a weather deck, sees the Russian helicopter, and darts back inside. "Maybe they thought it looked fun and wanted to join," he says.

The helicopter approaches. It circles the *Healy*, hovering 300 feet above the waterline. Then it passes again. And again. "Everyone's freaking out," an ensign named Patrick tells me.

"They were a little closer than I would have liked," Captain Schallip tells the complement later. The scene feels like it could have been plucked straight from the Cold War.

When the helicopter and boat disappear and the fog lifts, Captain Schallip gathers everyone in the science conference room for an all-hands meeting. "Today was the day we had planned for for a long time," Captain Schallip says. She recounts the events of the evening and early morning. It seems the event both rattles her and provides her with some relief. A mixture of gumption and pride fills her retelling. One of the science crew members asks whether this was normal, to see Russians in Norwegian waters. The Americans are visibly unnerved by the encounter, but this type of exchange is commonplace for the Norwegians.

"This is actually a normal procedure," Lieutenant Ocean, from the KV *Svalbard*, chimes in. "We do the same when they come down our coast. They're allowed to be here and work here."

WHILE OTHER NATIONS HAVE many icebreakers, the United States has precious few. They are not wholly necessary. "We simply do not have a similar demand in the U.S. Arctic by comparison, so while we certainly need to modernize and increase capacity, we should avoid creating any expectation that we need to match the Russians," one coast guard official tells me later, though even President Trump has called for a "fleet" of icebreakers since at least 2018. It is, however, a weakness when considering the projection of soft and hard power, more so if the Pentagon finds it impossible to build them at all. And the *Healy*'s hobbled appearance and incapabilities do not help America's reputation for strength in the arctic, be that militarily or scientifically. *Healy*'s encounter—the American crew's response—seems altogether of a different era, as though American capacity and ability in the arctic remains not hindered by its ambitions but by its lack of infrastructure and political will. But the United States has been here before. The confrontation, which, yes, occurred at Parry's farthest point, and, yes, at the point where Nansen lodged the *Fram*, was also the location of a skirmish between the Soviet Union and the USCGC *Northwind* some six decades before, during the height of Cold War hostilities.

In the summer of 1965, the *Northwind* deployed from Norway on a scientific mission into the Barents Sea, with the goal of returning home, across the top of Russia, through the Pacific Ocean. The U.S. Coast Guard had planned to plumb the Kara Sea to provide data to oceanographers, a not-so-different mission from the *Healy*'s own. The Soviets had back then published reports of their findings, but never shared the raw numbers— similar to their tactics today. Russia's presence in the arctic was such that the missing data concerned nearly half the arctic's waters. Without that information, operating in the shared north was like closing one's eyes and attempting to walk a tightrope.

Everything that happened that summer was eerily echoed through the *Healy*'s experience: the Soviets scheduled war games in the area, and several Soviet vessels and aircraft roamed the waters and airspace around the *Northwind*, including a guided missile frigate ("239") and a destroyer ("DD 020"). The *Northwind* commander sent an invitation to dinner through the ship's signal lamp. "I AM VERY SORRY BUT I CANNOT DO IT," came the Soviet response in Morse code from their own signal lamp. Back then, the

88

Soviet Union was protecting its polar seas. Back then, the Soviet Union had objected to the *Northwind* passing through its waters. The U.S. government bowed to that objection, and the *Northwind* returned by way of the Atlantic, having not completed its objective.

This time the U.S. Coast Guard completed its mission, but the *Healy* dared not freely roam through Russian waters to do so. Even as aircraft and vessels approached it, the only show of force the United States could muster was a flyover by a Poseidon spy plane, which arrived long after the Russians had departed. It was delayed because ice had formed on its wings, making it impossible to descend below the frigid cloud cover. By the time it arrives, the Russians, rulers of the arctic, are gone.

SCIENTISTS, RESEARCHERS, DIPLOMATS, SPIES

AT A HOTEL BAR north of Sør-Varanger, in northern Norway, on the border with Russia, a man introduces himself. (In the arctic, a drink and companionship are never very far.) The menu offers whole lobster for roughly $200. Drinks are cheaper and offer better fortification. We decide on a liquid dinner. The man's name is Zoran. He lives in Kirkenes. He has many questions—about politics, the arctic, my work. Who and what are my interests up here in the forgotten north? What is my mission? Who have I been speaking with? A nice fellow, charming even. The rest of the evening is a blur, save for Zoran and the seals poking their glossy snouts through the glassy stillness of the harbor, which is a motionless slate of gray beyond the picture windows.

A few weeks later, a journalist colleague messages me to say they, too, had met Zoran at the bar. The colleague wonders whether Zoran might be a spy. He asked a lot of questions, seemed to be cozy with foreigners, the colleague says. I say I do not think so. Then again, at only three kilometers from the Russian border, Kirkenes is and has historically been a great place if one is looking to construct a yarn, however flimsy, that would warrant a page in a John le Carré novel. It is a place where everyone is suspect.

"Everyone wants to know about spies," one Northern Brigade frontier commander tells me, rolling his eyes as he looks out over Kirkenes during military exercises, an increase in which has brought military units and reporters from across NATO member states. Those are the least of his concerns. A few hours later, I am in a rickety old pickup traveling the Russian border with a scientist who, with one hand on the wheel, uses his other hand to reach into his breast pocket. He pulls out a stiff sausage. "Meat banana," he says. "You want one?" I politely decline. A reindeer blocks our path and we slow down, the frantic steps of the lone animal on a backcountry trail at once somewhat comedic and melancholy. The scientist is Dr. Paul Eric Aspholm and he, along with his assistant and graduate student, Eline Hansen, is spending the day collecting samples across the Pasvik Valley. Dr. Aspholm is a researcher at the Norwegian Institute of Bioeconomy Research, studying ecosystems. His research across the arctic runs the gamut from the health of local pink salmon to the soil composition of makeshift firing ranges where the military tests its weapons.

While many scientists collect their research data through remote sensing in the arctic, Dr. Aspholm lives here. Dr. Aspholm, who punctuates most sentences with the word *super* and has the frenetic air of an over-caffeinated college professor, chases down the sausage with loud gulps of Coca-Cola. The two scientists are in the middle of a twelve-hour expedition, which began around ten this morning. Over the next half-day, we will visit more than a dozen confluences along the Paatsjoki River, taking water samples as they go and loading the samples into Styrofoam coolers that slide around in the bed of the truck.

The water here has changed since Russia introduced nonnative pink salmon into the arctic waters—a threat to the native Atlantic salmon. The *gorbusha*, as the salmon are called in Russian, have been a resource for that nation since millions were taken from Pacific Ocean tributaries and relocated to the Kola Peninsula in the 1950s. Already considered endangered, the Atlantic salmon cannot vie for the same resources against the pink salmon, which spawn and die and pollute the waters in which the Atlantic salmon swim and hatch. What results is a bloating, stinking, foul mess of dead Atlantic salmon carcasses all around the Paatsjoki River and its off-shoots. "There was once an ad about Norwegian water being the best," Dr. Aspholm tells me as we reach one of the river's tributaries, the Grense Jakob-selv. "Not anymore."

The water of the Grense Jakobselv slithers by in the shadow of a border guard station near two boundary markers. A blue sign warns of the international border with Russia. We will soon step into the waters anyway. This is not quite new territory for Dr. Aspholm, who visits the same stretches of the same rivers at the same times monthly, but it was also not that long ago that he had worked in Russia, at the Pasvik Nature Reserve. He had a special passport that allowed him the freedom of mobility to traverse the two countries. His job then was "running after bears and snapping off the ears of the bucks, counting the birds and the fish, and swimming in the rivers, working with the freshwater mussels." Paradise. But that was now a distant memory.

He stretches into waders and readies the telescopic pole he uses to collect water from the river. His primary interest on this day is how the ecosystems along the river, which eventually leads to the Barents Sea and out into the Arctic Ocean, have changed in the last decade. How will they continue to change? Who along the shorelines will be impacted? Upstream and to the east is Russia. What are they doing up there? How goes the stewardship of their land, which thanks to the river is shared? After all, on the Russian side sits Nikel, the mysterious home to one of the largest nickel-smelting plants in the world, where sulfur dioxide blanches the wilderness around town into a brown moonscape. "So what is interesting and why the high north is often in the international interest and especially important for Russians is that more than seventy percent of most resources are confined within the Barents region except for coal, which is in Ukraine and the southern part of Russia," Dr. Aspholm says between dips in the river.

He lists the numbers of military personnel on either side of the border, Russian and Norwegian, noting that they are similar—fewer than a thousand and mostly conscripts. The rest of the Russian forces have been "transported away" to Ukraine. But the guards here on the other side of the river have their purpose. It is important land, on both sides of the border, and for more reasons than simply sovereignty and science.

One of the projects Dr. Aspholm worked on more than two decades ago was a war game aptly called "Iron Curtain." "We were making scenarios, and actually a war between Russia and other southern European countries was one of the scenarios. After two years of war in Russia, they needed to expand to get enough resources" to continue their war. Was the tabletop resounding today? Russia did not win the exercise, but they did not exactly lose. Dr. Aspholm calls it "a type of stalemate." What they learned was that a conflict in the north could be spurred on from elsewhere, if a nation—in this case Russia—were to expand north in search of natural resources like those at Nikel (or, in the case of Greenland, cryolite, used to produce aluminum) or at the iron-ore mines of Swedish Lapland. "The same thing was in a scenario of climate change," Dr. Aspholm says at one point, straddling the river. "So climate change is pushing all Europe and Russia to the north, and the political will harshens because of resource exploitation and the needs for that exploitation."

Engines rumble in the distance. The border guard is on patrol. Sámi reindeer herders are also nearby, racing around their land, usually with a knife and *kuksa* (hand-carved wooden cup) hanging from their belts. Dr. Aspholm has a *kuksa*, too. He blends in well. The cup's varnish glistens in the sun as he looks up from his work to survey the area, mostly curious, like a dog whose ears perk at an unfamiliar sound.

"Interesting to see the changes," he says. "Not that I was spying, but just seeing."

"Everyone is a spy in their own right," I say.

"Yeah," he chortles. "You always observe."

EVER SINCE RUSSIA WAS largely shut out of Arctic Council meetings, scientists have struggled to find ways to continue their research collaborations with a major arctic stakeholder. Though Dr. Aspholm may have been joking, the effective weaponization of science is a real consequence of nation-states limiting the abilities of their academic and research institutions to cooperate with Russia. To be sure, there were authentic spies and acts of espionage ongoing, and the risk of more of both is real, but in the fog of war, it is difficult to discern who is doing what.

"We're not at peace," Jardar Gjørv, a researcher at the Arctic University of Norway, said one night over dinner in Tromsø, after I had returned from the border region. "We're at the first step to war." Gjørv works at the university's Grey Zone project, specializing in hybrid warfare research, founded by his wife Dr. Gunhild Hoogensen Gjørv, who tells me, "You can feel the pressure coming on" since 2014. The "U.S. is kind of ignoring the arctic, and that's just as well."

It may have seemed that way, even as Washington made small inroads on its own espionage mission in the arctic. In June 2022, congressional staff members were quietly discussing the first American outpost in the arctic since the Cold War. Back then, the U.S. government had an official presence in Tromsø—what it called an "information and cultural center." It was closed in 1966, and then immediately reopened as the U.S. Information Agency's Information Office. The Information Office was staffed by an American Foreign Service Officer. It closed again in 1994, after the end of the Cold War. Then on August 6, 2022, the U.S. Ambassador to Libya extolled the virtues of the outpost in an email to State Department staff members who were interested in setting up an American Presence Post, a sort of diplomatic mission, in the Norwegian Arctic. "We never should have closed it (for budget reasons in the mid-1990s) and I'm delighted we plan to reopen," Ambassador Richard B. Norland wrote in an email to the staff.

The State Department moved quickly. A lease agreement with the Fram Centre in Tromsø would have taken the agency too long; it needed an outpost in the arctic quickly. The State Department signed a Memorandum of Understanding instead. Espen Barth Eide, Norway's Minister of Foreign Affairs, gave his blessing for the mission in a letter dated February 24, 2023, and by August an "Arctic Watcher" was posted to the lone office, which by then had already been outfitted with shatter-resistant window film. The location was controversial, as the Fram Centre was dedicated to scientific research, not diplomatic affiliations. Still, the diplomatic mission would serve as a listening post in the High North, once considered NATO's Achilles' heel. It would monitor increasing Chinese interests in the arctic, identify vulnerabilities to foreign investment and resource competition in Svalbard, telegraph to Russia that Washington was watching, and offer America access to areas throughout Norway where China and Russia were increasingly active—specifically on Spitsbergen and in Kirkenes. A call with Senator Murkowski of Alaska was meant to reassure her of the State Department's commitment to the arctic, but any mention of increasing the department's partnerships with Alaska and Alaska Native communities was a footnote in the otherwise dozens of pages of internal communications about the new diplomatic mission.

While America slowly returned to the arctic, espionage, intelligence, and counterintelligence operations began to affect U.S. troops on deployment to the High North. Radars were jammed, GPS signals were spoofed, dangerous military maneuvers were avoided at the last minute. The Gjørvs and the Grey Zone project saw these attacks and more firsthand. José Giammaria, a Brazilian academic, arrived as a volunteer researcher for the program in December 2021. He was working on self-funded research toward finishing a master's thesis. One of his former colleagues described him as uninspiring, quiet, easily forgotten. "We went to dinner and it was unremarkable," Dr. Marc Lanteigne told me. "He was quiet" and very reluctant to talk politics. He did not use social media and he refused to join WhatsApp, restricting his digital communications to Telegram, the messaging app popular in former Soviet states. Dr. Lanteigne, who teaches international relations and China politics in Tromsø, is Canadian, and he had trusted—as did others at the university—the credentials and connections Giammaria had garnered over several years of studying in the North American Arctic, where he wrote about Russia's threat to global arctic security and expressed interest in northern maritime studies.

Trusting Giammaria was a mistake. The Norwegian Police Security Service arrested Giammaria, whose real name was Mikhail Valeryevich Mikushin, in

2022, for being in the country under a false identity. He was later identified as a senior Russian military intelligence officer working in the GRU, Russia's military intelligence agency. He was returned to Russia in a historic prisoner swap in August 2024, the largest since the Cold War and signaling Mikushin's and the arctic's importance to Russia's intelligence community.

Gunhild Gjørv later noted the irony of the incident: the Grey Zone project's main focus is hybrid threats against civilians. She also hinted at the flattery of it all. Russian officials were interested in her team's work. They must have stood out and done something right. America's interest, too, was evident in its decision to open its first foreign posting above the Arctic Circle, in Norway—a response to the ambient war and the increasing isolation between Russian and NATO-aligned scientists and diplomats.

Increased interest sowed distrust here as it had on Spitsbergen. Divisions grew between researchers who once closely collaborated with their Russian counterparts. Russian diplomats based in Norway and Sweden were expelled from those countries. American and European researchers were banned from traveling to the very regions they had studied for decades. One academic lost his post and tenure after attending a Russian arctic symposium.

Suspicions of espionage and infiltration came in two stages, in rapid succession. First, Chinese and Russian diplomats were expelled from Norway, Sweden, Finland, Iceland, and Denmark—a move extraordinary in its scope and scale. Then, infiltrations like that of Giammaria and increasing hybrid warfare attacks like critical infrastructure sabotage began shortly after the expulsions, extending well beyond Scandinavia.

INCIDENTS ACROSS NORWAY MADE America's greatest ally in the arctic seem an inflection point for conflict across the greater arctic. What cracked there would eventually crack elsewhere. But international responses varied. The Americans saw Norway as a training ground. The Russians saw it as a threat to their national security. The Chinese saw it as a toehold onto arctic politics and infrastructure and expeditions. The Swedish and Finnish saw it as a major conduit in supply and logistics chains, an ally that would be of the highest import not only during war but also during the limelight of peace. The Central Intelligence Agency had long noted that Norway's cooperation with America was paramount to any future successes in the north. Supporting Norway meant supporting American interests. This, as with anything out of Washington, began with the quiet opening of diplomatic missions like the one in Tromsø, then the appointment of an attendant of that greater mission of a more dominant role in the arctic. It was in this appointment that the questions of allegiances, intelligence gathering, and rites of passage came barreling home.

Dr. Mike Sfraga, the founding director of the Wilson Center's Polar Institute, a think tank in Washington, D.C., was nominated in 2021 to become the nation's inaugural Arctic Ambassador-at-Large. The position had its roots in the Obama administration, when Admiral Robert Papp Jr., a retired commandant of the U.S. Coast Guard, served as the Special Representative of the Arctic. However, Papp's role did not carry the same weight as Dr. Sfraga's newly established ambassadorship, which required Senate confirmation and came with a broader mandate.

The arctic ambassadorship was intended to "counter foreign malign influence" in the arctic, according to Senator James Elroy Risch (of Idaho). The American posture had changed during the confirmation process, as Russian and Chinese aggression across the region proved an increasingly malleable threat. During Dr. Sfraga's confirmation hearing in March 2024, Senator Risch emphasized that the position must focus on "national security challenges, economic opportunities, and the implications of U.S. foreign policy in the arctic." That included addressing issues like climate change, economic development, and Indigenous rights, but the scope and necessity of that position toward national security had changed.

When President Trump introduced the idea of purchasing Greenland (once during his first term, twice leading up to the start of his second), the statements seemed remarkably illiterate with respect to the history of America's feeble presence in the arctic and, furthermore, ignorant of how quickly that presence can become unwelcome. It also illustrated the stubborn myth that the arctic can be claimed by anyone under the right circumstances, that the region was somehow detached from international laws and norms, demonstrating a complete ignobleness toward ontological security in the region. Before Trump, the United States had on no fewer than three occasions (dating back to at least 1867) discussed the idea of purchasing Greenland (and Iceland), with the Kingdom of Denmark shutting down the idea each time. In 1975, C. L. Sulzberger, the *New York Times'* Pulitzer Prize–winning foreign correspondent and newspaper heir, wrote that "Greenland must be regarded as covered by" the Monroe Doctrine. The arctic island nation has remained in the American security perimeter firmly ever since. It is already an American partner in all the ways that matter.

Threatening takeovers in the arctic, suggesting the region is a place for outright military conflict, is what has historically deadened cooperation in these high latitudes. "It was a pretty dark message about the arctic, the expectations of conflict," one senior State Department official told me. The official was recalling when Trump's first secretary of state, Mike Pompeo, derided China and Russia in 2019 for their increasingly crude ambitions toward arctic dominance, which included funding infrastructure projects for dual-use military and scientific purposes or outright building military bases. It was around this time that Trump first sought to buy Greenland. "It failed to take account of what was going well and to convey a more optimistic, positive message. What you can change can be self-fulfilling," the State Department official said. The Arctic Council "was not a place where armed conflict would break out," the official told me. "It was a mutual place where rule would be followed." But those calculations had changed. "Our ambitions were that if there was a conflict between Russia and the U.S., it was more likely to start elsewhere and then spread to the arctic," the senior official said, adding that in all internal discussions the arctic had until recently seemed afar, elsewhere, over the horizon, and never at home, no matter who sat in the Oval Office. "The State Department is guilty of this: Automatically everyone thinks of Europe, not the American Arctic, Alaska."

Dr. Sfraga's confirmation dragged on for more than two years, in part because then-President Biden's reelection was uncertain (traditionally, the ambassadorship does not carry over to a new administration). For a few years it seemed like the delay was bureaucratic, a simple matter of a complex system working its course and wending its way toward confirmation. No one could dispute that Dr. Sfraga was a choice candidate; his long devotion to Alaska (he lives in Fairbanks, Alaska, and was raised in Brooklyn, New York) and arctic issues was easily apparent. Dr. Sfraga's expertise as a geologist, combined with his years of leadership at the Polar Institute, positioned him well to handle the dovetailing issues of climate change and arctic security. The position's mandate favored his experience: The arctic was once more a theater for strategic competition, with both Russia and China seeking to expand their influence in the region. But how he had gone about representing the institute and his past work in the arctic was not without scrutiny.

Dr. Sfraga traveled frequently, having embraced the role of a postulant diplomat since he became vice chancellor for university and student advancement at the University of Alaska Fairbanks in 2010, and then chairman of the U.S. Arctic Research Commission in 2021. His interest in national security, however, had increased in the several years prior to his nomination for arctic ambassador. One could in any given semester find him in Germany for the Munich Security Conference, in Finland for the Helsinki Security Forum, and in Poland for the Warsaw Security Forum, among other places. He skipped out on the smaller gatherings, like the Security & Defense on NATO's Northern Flank forum, likely because they did not offer a sufficiently grandiose platform on which to speak about his vision for a more unified arctic, a goal that he called, in the scholastic form of a professor in search of a pneumonic, the "Seven C's."

But by the time the confirmation hearings actually happened, tensions were higher, and suddenly Dr. Sfraga's experience working with Russia, China, or whomever else, became a weakness, with people like Senator Risch questioning whether he might pose an intelligence risk.

Senator Risch, the chair of the Senate Foreign Relations Committee, which was overseeing the appointment hearing, had reservations about Dr. Sfraga for all his globe-trotting and the failure to detail all his previous travels on his application for the position. "I rise today in opposition to the nomination of Michael Sfraga," Senator Risch began, before detailing what he called a pattern of behavior that demonstrated Dr. Sfraga was "not qualified" in matters of national security. "I believe he makes the situation worse. Based on his evasiveness during his vetting by the Foreign Relations Committee, I believe Dr. Sfraga could pose a counterintelligence and foreign malign influence threat to our nation," Senator Risch said, before adding, "I don't say that lightly."

Senator Risch laid out a compelling case for postponing the appointment in favor of a more thorough vetting process, despite the urgency he recognized in the need for a diplomat dedicated to the region. "Dr. Sfraga was asked repeatedly to provide his foreign travel, his foreign contacts, and his appearances on panels to the vetting personnel, and he failed to be open and transparent. He updated his file four times—maybe a new record—each time only after being confronted with additional information he tried to conceal. For instance, while at the University of Alaska Fairbanks, Dr. Sfraga negotiated no less than twenty-seven memorandums of understanding (MOUs) with Chinese academic institutions tied to China's intelligence services. Only after confronting him did Dr. Sfraga admit that he negotiated these MOUs. One in particular was with a Chinese university with ties to Chinese intelligence services, and the MOU gave the Chinese access to the university's IT systems, exposing it to substantial cyber threats. On Russia, Dr. Sfraga failed to disclose a panel he spoke on in November 2021. Transneft, a sanctioned Russian state-owned energy firm, sponsored this conference."

Senator Risch continued: "I really think that he is naive, at best, as far as dealing with Russia and China. And, in his defense, the entire academic community, for that matter, shares this naivety when compared to our national security agencies." He pointed to past articles written by Dr. Sfraga about the need for greater cooperation with Russia and China. Such cooperation had become suppositional in preceding years, despite its status as a founding pillar of the Arctic Council and post–Cold War advancements in diplomacy and climate initiatives. Most would agree that cooperation was necessary, as did Senator Risch. But that did not stop the senator from taking the unusual step of formally requesting that the Federal Bureau of Investigation conduct a supplementary background check on Dr. Sfraga—a request that was ultimately denied by the Biden administration.

Senator Murkowski was Dr. Sfraga's biggest champion on Capitol Hill. "He has been very clear about making sure that U.S. interests in the arctic are protected and defended—against Russia, against China, and anybody else that would encroach on our sovereignty," Senator Murkowski said during the hearing. The Senate vote passed 55 to 36, with Democrats in the majority.

At first blush, the hearings seemed to revitalize a caustically hawkish arctic discourse: Everything was crumbling, nowhere was safe, and the only way to confront any problem was with one extreme response or another; that in the arctic there was only a zero-sum game to be won. Never compromise. It was the exact thing Dr. Sfraga sought—greater cooperation, however leery one may be of their partners—while traipsing around the arctic to visit nations both friendly and hostile. Dr. Sfraga had acknowledged that this was simply a new reality across the arctic. Even the White House had acknowledged in its most recent National Security Strategy that the arctic and its needs were being met, in part, through cooperative agreements, like a pact to collaborate on icebreaker construction. The arctic, for Dr. Sfraga, for all his past posts in the scientific community, was not just about climate change. He understood that the Department of Defense needed to better understand the northern environment and that it needed to operate there in a different way from how it had during the Cold War—and how it did today. Tensions had crescendoed to the point where Dr. Sfraga's history of working with Russian and Chinese scientists and diplomats had nearly sank his nomination. Finally, those did not sink it; tradition and matters of decorum would.

Ambassadors traditionally opt to step down when a new president is elected. One source, recalling internal discussions among arctic-focused colleagues in the field of security research, told me that "the incoming administration does not want to keep him and likely won't replace him, especially given that they are planning major cuts at the State Department." On January 20, 2025, Dr. Sfraga tendered his resignation and the post remained unfilled.

There was one thing I took away from this saga: No matter where in the arctic one resided or governed, the entire circumpolar north shared an unwavering affinity for the way things were, and held a cautious eye toward what they still may become. Lacking affordable energy, lacking infrastructure, similarities in language, and access to critical minerals and materials, plus a continual brain drain—these issues were universal. Viewed in this light, Dr. Sfraga's travels, as with Dr. Aspholm's river expedition, were less foreign than travels to one's own backyard. Are we naive, as Senator Risch put it, to trust that what benefits one nation in the north benefits all arctic nations?

To answer that question, one need only step up to a bar, or wade through an arctic river along hostile foreign territory, to understand what cannot be bridged over diplomatic distances. To interact with is to learn, to gain an advantage, to destigmatize.

The benefit to all is now undermined by Trump's continuing designs on Greenland. What could have been an opportunity to rethink the way America bands together with allies and foes alike to regain its erstwhile strength in the north is now replaced by the severing of ties and abandonment of cooperation in favor of opaque desires and isolationism.

"There are other nations questioning us," Dr. Sfraga said at an Alaska legislative hearing in April 2025. "Every one of our allies in the north and elsewhere, they're questioning how committed we are."

THE LAW OF
THE REALM

NICHOLAI'S CLASSMATES CALLED HIM "Chocolate" in elementary school, a fact he recalls with bitterness and without explanation. Rotund, with a smooth jaw and demure eyes, Nicholai, thirty-two when we meet, works as an electrician across Greenland, installing fire and security alarm systems for a Danish company. He is originally from Ivittuut, an abandoned mining town on Greenland's southwestern coast. His favorite foods are cured narwhal and polar bear, though he rarely has the money or opportunity to eat them. When he speaks with other Inuits, the language is strained but comprehensible. He spends most of his time speaking Danish.

We wait to board our plane in Reykjavík. We are heading north, but Nicholai's trip has been much longer. Having traveled from Nuuk, the quickest way to the remote coastal village of Ittoqqortoormiit, in northeast Greenland, is to fly through Copenhagen, then Iceland, before boarding a Super King Air B200 for the final leg of his thirty-six-hour journey. He transited two other countries just to get back home. The circuitous route is emblematic of Greenland itself: a country geographically closer to North America but culturally and bureaucratically tethered to Europe and Denmark, thousands of miles away.

On the plane, Nicholai takes a photo of downtown Reykjavík, covered in a fine, fresh dusting like a village in a snow globe. He looks around the plane. It is small and packed tight. Up front are the pilot and co-pilot. Behind them are seats filled with mail and cargo. Nicholai is traveling with his boss, an air traffic controller, and myself, all bound for Constable Point, the only airport connecting the village of Ittoqqortoormiit to the rest of the world. Even then the settlement is still an hour-long snow machine ride, or a fifteen-minute helicopter flight, away. Tourists rarely visit. If one is not fond of a desolate landscape surrounded by white peaks, endless wind, and few amenities, there is very little to "see" and "do."

At the entry of the waiting lounge at Constable Point, two degrees north of the Arctic Circle, there is a small box of blue booties to protect the carpets from snow- and mud-covered boots. Signs across the room warn against the use of excessive perfumes or colognes: Greenlanders are largely allergic. They are also leery of anything manufactured, anything from outside intruding on the natural. A price list on the wall gives the cost of room and board. It is not uncommon for those passing through Constable Point, or anywhere else in Greenland, to be delayed for days. Airline tickets are in part so costly because of this: the airline covers the housing for passengers stranded due to unpredictable inclement weather. For all the struggle and difficulties facing Greenlanders today, from a messy and slow-moving search for independence to atrocities committed against Inuits for Western gain, at least they are not subjected to the onslaught of fragrances and aftershaves typical of airport duty-free zones.

Peter, the airport manager, greets the air traffic controller and the rest of our group. The helicopter transfer has been canceled due to freezing fog. Dinner will be served in two hours. Nicholai shrugs and hauls his bags and equipment outside, dragging them past a series of airport hangars to a small railroad building painted red. In 2016, a Hilton logo had hung above the entrance, placed there by employees of the airport. Below the logo was a minus sign adjacent one star.

The aerodrome beacon sweeps over his window every three seconds. He pulls the blinds shut and sits in darkness until dinner—curried herring, thinly sliced roast beef, dark rye bread. A Danish spread. He is home, but it feels like being in Denmark.

THE LAW OF THE REALM

GREENLAND HAS BEEN UNDER Danish rule for more than 400 years. For Inuit like Nicholai, distant representation from Copenhagen can feel like being back in grade school: relegated to second-class status in their own land. Since 2008, the Greenlandic government has taken steps toward independence. Though it has the right to declare full sovereignty at any time, Greenland relies on an annual $511 million block grant from Denmark to sustain its economy. While the Greenlandic parliament governs education and health care, foreign affairs and the judiciary remain under Danish control. And even within the so-called autonomous realms, which include the Faroe Islands, Danish influence is felt.

When a Chinese firm proposed expanding Greenland's lone international airport to accommodate more direct flights, Copenhagen shut down the investment and oversaw the project themselves. Ostensibly, it was a national security concern. Nuuk politicians viewed it as a diplomatic setback—a decision expected to fall under the purview of Greenland's semiautonomous government that was overridden by Danish fiat. The tension revealed the fragility of Greenland's ambitions of autonomy.

The philosophical divide between the two nations is embodied in Janteloven—the Law of Jante—a Scandinavian social code that values conformity and discourages individualism. Taught to Danish children as early as preschool, its tenets include "You're not to think you are smarter than we are," and "You're not to imagine yourself better than we are." But in Greenland, where cultural resilience has often depended on standing apart, not blending in, Janteloven is viewed with skepticism if not outright rejection.

"Greenland is a black hole," said Karl-Peter, a maintenance worker at the airport. "Nuuk is the center, destroying the country."

That view is widespread. Though "home rule" was granted in 1979, and further expanded in 2009 with a "self-rule" agreement that recognized Greenlanders as a people with the right to self-determination under international law, many Greenlanders feel little has changed. Financially, the country remains dependent on Denmark—to the tune of roughly $11,000 per citizen each year. More than 70 percent of the population supports independence, but Danish and Inuit cultures remain entangled.

Nicholai is emblematic of this entanglement. He teaches his children Danish, depends on Danish health care, banks, and infrastructure. Yet he lives and works in a country where self-determination is a national aspiration. When he arrives at Constable Point, the majority of the staff are from Jutland, in western Denmark. Their plaintive views on Greenlandic independence stand in contrast to their very presence: temporary workers housed in a temporary place, maintaining a semipermanent hold.

Ittoqqortoormiit is not a historical Inuit settlement. It was established in the 1920s under Danish law, part of a broader effort to populate and secure Greenland's coasts during the early twentieth century and in the wake of two world wars. In the 1940s, the Danish government forcibly relocated communities to places like Ittoqqortoormiit. The aim was both geopolitical and strategic. The people have lived largely unsupported since, surviving against odds that have worsened with time. Climate change has disrupted traditional hunting cycles. Narwhals and seals no longer migrate as before; ice sheets are receding. In 2021, there were no narwhal to hunt. That year, the village's primary income source disappeared.

Before the climate changed, the Danish government had already imposed hunting restrictions and export bans on seal, polar bear, and whale exports. The villagers could sell only domestically, making them reliant on Danish imports and policies from Nuuk. It immobilized their culture. "We managed to live with others, hunting food," said Karline Hammeken, a resident of Itoqqortoormiit. "But not anymore." In summer, only two cargo ships reach the village, and cruise ships rarely call anymore due to environmental regulations from Nuuk—892 miles away. "For us hunters," said Nunduu Lorentzen, "this is our gold."

INSIDE THE TERMINAL OVER the ensuing days, Nicholai tinkers with the airport's electrical systems, climbing ladders and opening connection boxes, staring into blinking lights and unsophisticated security systems that say only whether an alarm is "active" or "ready," if there is an "error" or if the system is in need of "reset." The days stretch. The helicopter remains grounded.

I set about the air station, hopping from terminal to maintenance shed, and find Karl-Peter getting ready to paint the floor near the generator room. We grab brushes and paint and begin rolling out a messy coating of matte gray. The room, over the course of an hour, begins to look brand new, the floors now lacking the wear and tear of boot scuffs and chipped wooden planks. "I'm against [the independence movement]. I love Denmark. We get all the money from Denmark," Karl-Peter says as we watch the paint dry. "Imagine I give you money and I say, 'I will live on my own, fuck you, give me that money.' It's double moral." Karl-Peter moves a large hose out of the way. It is preventing us from painting the remaining corner. "They're still fighting. Living in the past."

Karl-Peter promises me some muktuk, or narwhal skin and blubber, after dinner, and we arrange to meet in the common area of the "hotel" later. When we meet, Nicholai is already there, and we sip Coca-Cola while eating muktuk topped with a Danish seasoning. The men are equally dark in their viewpoints on independence. They are on warring sides. On one, Karl-Peter, who is happy with what he has and treasures his mobility across Denmark; on the other, despite his dream of taking over the Danish company he works for, there is Nicholai, who doubts Greenland will ever achieve independence. Karl-Peter sees the infrastructure. Nicholai sees the inequality. He sees that sovereignty alone won't rewrite history or wire the country for success. And still, he believes in something else. Maybe not independence, but dignity.

GREENLAND'S STRUGGLE FOR INDEPENDENCE is not only cultural or economic, it is strategic. In 2025, when President Trump declared that "we need Greenland for national security purposes," he echoed a lineage of imperial designs. But Greenland is already integral to U.S. defense—

through Denmark, a NATO partner, and Pituffik Space Base, the American outpost at the northern edge of the island, which was enabled by one renegade diplomat.

In 1941, Henrik Kauffmann, Denmark's ambassador to the U.S., defied his Nazi-occupied government. He declared that Denmark was under duress and therefore unable to govern its overseas colonies. Acting unilaterally, he signed an agreement "In the Name of the King," transferring control of Greenland to the United States. He used Denmark's U.S.-based gold reserves to fund Danish embassies across the Americas and Middle East. The move allowed the Allies to build arctic air bases and weather stations, helping to shape the outcome of the Second World War.

Kauffmann's agreement laid the foundation for what would become a permanent U.S. military presence. Greenland's role in Western defense is not new. What is new is the language of ownership and transaction. "We don't want America to play an oversized role," a Danish diplomat told me in Copenhagen. "But we want all of our allies to bring Greenland to the West." Today, interest in Greenland stems not only from geography but from geology: rare earth elements, minerals, uranium. Chinese-backed projects like the Kuannersuit and Isua mines were eventually shuttered. American interest crowded them out. "Over time, we've seen U.S. interest phase China out," said Rasmus Leander Nielsen, head of the Nasiffik Center in Nuuk.

A more forceful push by Trump's State Department could unhinge the legacy of American cooperation and alignment with Greenland in Denmark. Trump's renewed interest in Greenland was met with firm resistance from Danish Prime Minister Mette Frederiksen, who emphasized that Greenland is not for sale and called for a united European response to any U.S. pressure. The Trump administration planned to offer financial incentives to Greenlanders, proposing annual payments of $10,000 per resident to replace Denmark's annual subsidy. Plans were announced to use social media to mimic the tactics of authoritarian countries seeking to sway the local populace. Vice President J. D. Vance visited Greenland to assert that U.S. protection would better serve Greenland's security needs, especially against potential threats from Russia and China, though the vice president failed to note that Greenland's security needs were already served by America. The Greenlandic public and Danish officials so roundly shunned the idea of a visit from anyone tied to Trump that officials could only safely visit Pituffik to give their stump speeches.

The flurry of activity in and around Greenland reminded me that the lesson of Janteloven, which the Danish failed to heed and which has cost them dearly in Greenland, was also being ignored by American officials. History or even a modest understanding of the ways by which nations' interests dovetail today was never Trump's strong suit. A real and very dangerous naivety persists in his approach to Greenland, and the arctic broadly.

At Inatsisartut, Greenland's parliament, I meet with Ole Aggo Markussen, party secretary of Siumut. He argues that with direct international flights and expanded tourism, the math of independence would change. But others in Nuuk express caution. One Danish official tells me Greenlanders "don't want independence. They want benefits. They don't stick to their promises."

Some researchers in Nuuk are proposing an entirely new framework: to replace the Danish Law of Jante with the "Inuit Myth"—a worldview that centers on ecological stewardship and Indigenous self-governance. It is a radical rethinking of arctic sovereignty, one that reverberates beyond Greenland to Alaska, Canada, Russia. For now, many I meet with across the country express hope and frustration, not least of all with Denmark's inability to defend Greenland. In Ilulissat, a picturesque town framed by glaciers and overlooking a channel dappled with icebergs and ice floes, a vacation town for foreigners and Danish alike, where potato confit with crispy cod can be had for a steal, I meet Lars Erik Gabrielsen, who proudly takes me for a tour of the homes he has built over the years. Rarely are they for his countrymen, but rather for outsiders. "They're cheating us," Gabrielsen says of the Danish as we drive through town. "They expect they can own us." The same could now be said of America.

There may be some hope of better relations between Danes and Greenlanders, between Greenland and the U.S., as younger generations of Danes who frequently come to the island nation for seasonal work learn more about the effects of their nation here. Rare is it that a ruling country's citizens take more than just a paycheck from their foreign colonies, and in this case they also seem to take with them a newfound respect for their impacts here.

"I feel really embarrassed about it," one Danish woman working as a server at Naleraq, a bar and restaurant in Ilulissat, tells me. "We treat them so awfully."

GREENLAND IS NOT SEEKING to be bought. Nor is it looking to be rescued. It is seeking dignity, cooperation, and respect. Nicholai, perhaps more than most I came to know across Greenland, embodies this contradiction. He is tied to Denmark, dependent on its systems, yet skeptical of its promises. And still he hopes. Greenlanders believe in a future governed not by Denmark's sameness but by their nation's difference. The Inuit myth.

At a town square in downtown Nuuk, I happen upon Nicholai a month after meeting him in the country's East. He is on a break and we embrace. We stand in the cold. He tells me the weather at Constable Point cleared eventually and that he made it to Ittoqqortoormiit. He visited homes, checked wiring, listened to the stories of those who lived through resettlement, though he knew those tales already. After a week, he returned to the patchwork itinerary that brought him home: Ittoqqortoormiit to Constable Point to Reykjavík to Copenhagen to Kangerlussuaq to Nuuk.

Nicholai updates me on his partner and his children and his dream of taking over the electrician business. He reiterates a feeling prevalent even in the faint light of promised independence as it shone against the realities of interdependence and the machinations of foreign nations. "You never get higher than the Danes," he says, noting that to look elsewhere for support and defense, if only for respect, seems like the nation's only option. With the Danes, "we're always lower," he says. Though under the defense umbrella of NATO and the United States, Greenland is willing to accept protection from whomever is willing to invest in its future. Not those who threaten to invade it. But those who lift them up.

"THEN I'D BE THE ONE SENT TO GO FIND THEM"

THE VICTUALS OF A Swedish winter—alcohol-free rosé, gingerbread snaps—lay strewn across the crowded kitchenette counter along with plates and other sweets, like salt-covered black licorice and a bag of warm *lussekatt*, sweet and spongy saffron buns served during the holidays. Snowdrifts hug the officers' quarters moonlighting as a medic station. The raking light, with its gauzy swirl of blue and pink, reaches Kalixfors, a shooting range and military base in Sweden's northernmost county. The county around Kalixfors is roughly a quarter of Sweden's total landmass. The base itself is a mile south from the city of Kiruna, in Swedish Lapland, ninety miles north of the Arctic Circle. In several barracks nearby, the conscripts sleep. Their train south to Stockholm will depart later this week, just as first light sprinkles through the birchwood. (Trains are cheaper than flying, and it is not uncommon for a jet-fuel truck in Stockholm to freeze before it can supply an aircraft for takeoff, causing delayed or even canceled flights.) The cold does not dissipate by midday, and the December light yawning across the barren fields awakens nothing with its weak amber haze. Outside, temperatures dip below minus 4 degrees Fahrenheit.

The conscripts, as part of a mandatory one-year service, have spent the last week running drills in this weather. They were choked with tear gas, practiced overland navigation with topographical maps made indecipherable by the thick snow masking the terrain, and maneuvered heavy-tracked articulated vehicles. They skied, a lot. Now they are allowed to sleep in. After holiday break, they will return in January and February to continue their training. Recently there's been a greater sense among the conscripts that this is not simply a civic service performed for a few months before heading off to college; increasingly it feels like real-world preparation, rather than a mere ceremonial commitment to their nation.

"It feels different than before," a young blond conscript from Stockholm says during one of the drills. "It feels more real. We talk about it among friends." Traditionally Sweden, and its neighbor to the east, Finland, have been largely neutral in times of war, unaligned and independent. But both nations applied for membership in NATO after Russia's invasion of Ukraine. The two formal applications, which required unanimous support from all NATO members, took separate tracks. Finland was quickly accepted. Sweden's application hit a diplomatic snag; Turkey held up the process for months, seeking to force Sweden to amend its constitution and anti-terrorism laws, which it saw as enabling the safe harbor of Kurdish rebels considered terrorists by Turkey. Sweden acquiesced to most of those changes.

Just a few weeks before the conscripts arrived at Kalixfors, the Turkish Parliament's foreign affairs committee finally passed the measure to the full assembly, the penultimate step toward Sweden's accession into NATO. In those few weeks, conscription applications (generally a mandatory service, with some exceptions) saw a 1,000 percent increase—evidence that more and more young people across the nation felt their security threatened and wanted to join the Swedish Armed Forces. Many of them would start their training at Kalixfors, cold weather being the foundation of Sweden's military survival training.

For now, the television in the officers' lounge overlooking a half-ring of couches and a coffee table is turned to local news. A local news broadcaster talks about young animal-rights activists who set free a tractor trailer full of chickens and vandalized the eighteen-wheeler in the process. "Funny," Sergeant Oscar Vidlund, one of the drill instructors, says. The next segment reports on a large indoor swimming pool complex coming to Kiruna, a project already facing cost overruns. "Ridiculous," Lieutenant Adam Hagman, another one of the drill instructors, says. Both drill instructors sip heartily on large mugs of coffee that block their entire face as they tilt back their heads. Both sport unkempt mustaches. Both are part of the Norrbotten Regiment (Lapplands jägarregemente)—one of Sweden's largest. They are tasked with the development and training of operational units specialized in winter warfare for the regional brigade.

Now, for international news. Secretary of State Antony Blinken today signs a Defense Cooperation Agreement with Sweden. The atmosphere in the room cools slightly. Neither Sergeant Vidlund nor Lieutenant Hagman have any more quips. Sergeant Vidlund stands to fill his mug. I grab a gingerbread snap and sit quietly between the two men dressed in military fatigues, feeling very much like these are the early days of war, that I am the foreigner who is to blame.

"Feels like the U.S. is capitalizing on the war in Ukraine after the Cold War, since Finland and Sweden are tired of Russia," Sergeant Vidlund says. "It was strategically inept to ignore this region for so long, because if there is a war, it'll be fought here." America had awakened to its vulnerabilities in the north because of the war in Ukraine, but it also had used those vulnerabilities to goad friendly nations into improving their own national defenses, giving over access to the Americans for weapons and munitions storage and troop deployments. Once-neutral countries were taking sides as their military ranks swelled. This would, in the event of a northern conflict, broadly benefit America. Washington would need to devote less of its own resources to defending its allies.

Another sip of coffee for both. Their expressions belie any emotion.

Lieutenant Hagman casually flicks his wrist toward the window, where outside the conscripts and the base surround us. "When Russia mobilizes on the border, these are the units who will be called up first."

SWEDEN'S RELATIONSHIP WITH RUSSIA has long been one of suspicion and conflict. From the Great Northern War in the eighteenth century, when Sweden lost territories to Russia, to tensions over Baltic Sea dominance during the Cold War, the two nations have danced a cautious waltz of proximity and rivalry. Sweden's neutrality policy, while serving as a shield, did little to ease the geopolitical tension that hung for centuries over the Nordic region, particularly in its vulnerable northern territories. With the end of the Cold War, the temperature of the disagreements between Stockholm and Moscow seemed to cool, but beneath the surface, the seeds of mistrust remained insulated.

In recent years, this latent suspicion has reignited. Sweden has increased its military spending by 34 percent since Russia invaded Ukraine. It has also integrated more closely with NATO, eventually joining the alliance, participating in exercises, and opening its bases to foreign military personnel and equipment. The shift in military strategy reflects a deeper fear: that Russia's gray-zone tactics are already unfolding and proving effective at destabilizing Sweden's most remote and vulnerable regions here, in the high north of Swedish Lapland.

The vast, frigid wilderness of Kiruna and its surroundings are home to valuable natural resources, as Dr. Paul Eric Aspholm's war game had shown, from iron ore to rare minerals, which are critical to both Swedish and global industries (the European Space Agency's Esrange Space Center satellite launch site, the Stockholm Teleport Station in Ågesta, are both here, to say nothing of Swedish Lapland's 21-million-euro annual Christmas business) and provide some 80 percent of Europe's iron. It is a straight shot from the ice-free port of Narvik in Norway through Kiruna to Finland and into Russia, making Kiruna an ideal redoubt or staging ground for Western troops. Kiruna's location above the Arctic Circle has made it a target of strategic interest for years, but Russia's recent activities have raised the stakes. Since its invasion of Ukraine, hybrid warfare, which combines conventional military tactics with cyberattacks, disinformation campaigns, and economic pressure, has become a hallmark of Russian strategy the world over. Sweden and its Nordic neighbors have increasingly found themselves in the crosshairs of these tactics. Such activities typically occur below the threshold of open conflict, but are aimed at destabilizing a nation's political, military, or economic systems. The rail line carrying iron ore through the Kiruna region, tracing that same route from Narvik that Western troops might take during a northern conflict, has suffered derailments on no fewer than four occasions in as many months, generally when joint military exercises are underway nearby. Military experts and intelligence reports suggest that Russia is probing Sweden's defenses. This would make sense, because Sweden's NATO accession was expected to facilitate a deeper integration of its northern, arctic-space infrastructure with commercial broadband, intelligence gathering, and combat operations. Nevertheless, these hybrid attacks remain deniable on Russia's part, parcels of a broader campaign to destabilize Swedish interests without crossing the line into overt conflict.

Russia has not been alone in its efforts to undermine European arctic regions. A year or so after China broke ground on the China–Iceland Arctic Science Observatory in Kárhóll in 2016, the People's Republic of China began construction on the China Remote Sensing Satellite North Polar Ground Station in, of all places, Kiruna. It may be used to closely monitor NATO activities as the two nations team up against the West.

Hybrid, or gray-zone, warfare—which over time or rather quickly can destabilize a nation into political and social cannibalism—has emerged as a key tool in Russia's arsenal. By targeting Kiruna and the global arctic, Moscow and Beijing are not only testing Sweden's ability to protect its northern frontier—and its allies' willingness to defend it—but also sending a signal about its broader intentions in the arctic. The casus belli, should it emerge, will likely not be a dramatic invasion but, rather, a slow unraveling of security through these insidious tactics, making the high north of Swedish Lapland a geopolitical flashpoint. As a newly acceded member of NATO, it is also a region that, should it be called to war, all alliance members would be called to defend under Article 5, which stipulates that an attack against one nation is an attack against all.

Like the United States, Sweden has historically pulled back on investments in its northern territories, particularly in infrastructure and defense. Both nations initially regarded the north as a remote and less strategically significant area than more populated regions. In Sweden, the vast and sparsely populated north, including Kiruna and greater Lapland, saw limited defense investments after the Cold War, mirroring Washington's slow approach to arctic military developments. But with the defense cooperation agreement announced by Secretary Blinken on the television this morning, U.S. troops will now have access to seventeen Swedish military bases. Kalixfors is one of them.

THE LATEST ADDITION TO the Kalixfors training ground is a Russian helicopter fuselage. For training unspecified. That's all the instructors say, adding winks and nods. The gray metal body of the helo reflects the pink watercolor sky as a platoon of conscripts strap on skis and, in two single-file lines hugging either side of a road, set out. Tear gas from a nearby exercise wafts across the lines. Using sets of spindly metal poles, the conscripts push forward.

Proficiency on skis was not the only point of the five-kilometer mid-morning excursion. One of the biggest threats posed by cold-weather warfare is thermal regulation, when the mind slows so the body can focus its heating efforts elsewhere. Excessive sweating also dampens clothing, which then of course freezes. Even if the material of the clothing aids in the wicking away of some of that sweat, any moisture near the body can make a minus 4 degree Fahrenheit wind chill feel more like minus 76 degrees Fahrenheit. Keeping dry is the conscript's primary objective. So, the conscripts get a quick lesson in wardrobe management. The importance of this is made clear by the basic Swedish cold-weather uniform: two simple layers beneath a third thin layer that protects against snow and rain. After a few kilometers, the instructor leading the group, Lieutenant Emelie Lilja, commands everyone to head into the brush and to swap their wet layers for the dry clothes they have kept in their rucksacks.

The soldiers strip off wet socks and wedge them between their bodies and waistbands to dry during the next portion of the trek. Sweat is not the only killer; apathy is also a threat. In this weather, the loss of determination to stay dry and warm can quickly lead to death. A soldier calls from the thicket of birch and asks Lieutenant Lilja if they can build a fire or light their denatured alcohol stoves for warm water. Lieutenant Lilja is standing on the road in nothing but a black training bra as she changes undershirts. She says no. They need to keep moving, not to stop, to rely on themselves for warmth. Trust the gear, the basics, not the luxuries.

She tells me that for a majority of the conscripts the cold weather is new. Increasingly warmer temperatures in the south of Sweden, where many of the conscripts are from, mean that few have been raised skiing and camping in cold weather, as past generations had. The skis are equally important; snowshoes drag, slowing troop movements. The learning curve is steep, but skiing is the best transport, whether under leg power or being pulled behind a tracked articulated vehicle. They must learn to trust the efficacy of their training and gear, Lieutenant Lilja says. "If you choose one climate to learn in, this is the one. You can then handle anything."

That notion has caught on among allies across the north. Sweden's highest military commander long recognized Kalixfors as a natural place to train international units. American military personnel from the U.S. Special Operations Command Europe and the 10th Special Forces Group based out of Fort Carson, Colorado, have trained here in the past. But Swedish instructors and military personnel told me they were not sure those visiting soldiers ever really learned what was being taught here. One spring, with snowpack still on the ground, several American soldiers trekked out on skis for an exercise. They went shirtless and returned within an hour, scorched by sunburn, despite the mercury dipping into negative digits. They had failed to anticipate the power of the sun's reflection off the snow. In a private discussion among U.S. Marines, many of whom have taken turns in four-month deployments to northern Norway, they joked that they did not need to ski and that they lacked the will to learn when they could instead throw hand grenades.

The arctic is perhaps the only place on earth where the environment itself can be weaponized, whether it's electromagnetic warfare, communications jamming, or ICBMs. Perhaps the Americans had merely forgotten this and instead treated their training as a ski holiday.

As we make our way from the fringes of the woodland back to the base, where the conscripts will eat outside, served lunch from tall thermoses of soup ladled quickly—can't let the soup freeze—into their metal lunch pails, one instructor turns to me and says, "The cold doesn't discriminate."

In Kiruna, a region with a bloody history still fresh on the mind, there were renewed assessments of what happens next in the borderlands. The regimental offices of the Norrbotten Regiment are located in a building on which snow clings to the mortar between the bricks, a facade cast in the yellowish fans of half-moon lamps outside. Inside, a map of the Scandinavian Arctic was spread out on a large wooden table. The Defense Cooperation Agreement, paired with Sweden's accession to the NATO alliance, opened an entirely new borderlands for staging potential operations into Russia, and for defending against foreign incursions. "On NATO's eastern flank, on the Russian border, the 'pre-war' era has begun," one French newspaper headline declares.

Lieutenant Colonel Martin Myhrberg Andersson, the chief of the Norrbotten Regiment, spreads his hands across the map, occasionally pointing to Narvik, the Norwegian port where NATO and American supplies could come ashore and spread eastward, fast, across the arctic regions of Norway, Sweden, and Finland, reaching the Baltic Sea to supply and assist Estonia, Lithuania, Latvia, and the other Baltic members of NATO. He waves his hand across Scandinavia and says, "From a military point of view, it's more or less the same country. The planning has been ongoing for a couple of years. The whole security picture around us is much, much worse now than just two years ago. . . . It's accelerating."

Like their American counterparts, the northern parts of Sweden have been largely left undefended. While escorting a delegation of American commanders before our meeting, Lieutenant Colonel Andersson recalls standing atop the local ski slope, whose shadow stretches across Kiruna's downtown. From there, on a clear day, you can see the airport, the railroad, all the roads and the narrow terrain, nearly all the way to Narvik, 108 miles away by road. "One of the officers asked me how many anti-aircraft battalions we are defending this area with." He shook his head. None. "We had a twenty-year time-out—didn't understand what's going on around us," Lieutenant Colonel Andersson says. He irons the map with the palms of his hands. "We have actually neglected the north for a long time. Making this happen with Finland, Sweden, Norway, and the United States, and after that in NATO, is very important for our security up North."

The Americans come and train more regularly now, relying on their local partners like the Home Guard, a reservist detachment that lives year-round in Swedish Lapland to guide their exercises. The American forces come to learn, but the knowledge is rarely shared and has to be retaught every year. "I'm sure that some of them are correctly trained and have experience, but some of them did not. Some of them were great and really understood. And some of them couldn't handle it at all. They of course have an agenda of their own."

AFTER THE NEWS BROADCAST and a short clip of statements from the Swedish prime minister and Secretary Blinken, Lieutenant Hagman switches the channel. He lands on a reality television show on arctic back-country living. Sergeant Vidlund and Lieutenant Hagman take turns pointing out the failures of the cast members to properly tie down their tents with skis ("Those are going to snap") and the way they casually toss their gloves onto the snow ("Well, there goes your gloves!"). Master Sergeant Jan Ejeklint enters the room, and the instructors depart to see the conscripts off to their trains.

Sergeant Ejeklint has seen the American forces, has helped them train. He has not been convinced that what it takes to survive in the north has sunk in among the U.S. troops, though he has hopes they will continue training and visiting the European Arctic to learn alongside their newest NATO partners. In the past five years, there have been more and bigger exercises in the arctic regions of Europe involving American troops. In 2024, the largest cold-weather military exercise since the Cold War took place across the northern regions of Norway, Sweden, and Finland. The U.S. Army, Air Force, and Marine Corps attended. Special Operations Forces from the U.S. Navy and other branches held their own exercises later that year. The Americans, many of whom are undergoing cold-weather training for the first time on these exercises, often need a bit more time, a bit more patience, a bit more guidance than those who feel a war is already at their doorstep: while the American Arctic sleeps, or fixes its broken icebreakers, the European Arctic prepares for war.

So it was with a bit of reluctance and consternation on the day after the conscripts finish their training that Sergeant Ejeklint, who I now, with the conscripts gone, can call Jan, meets me outside the barracks in the darkness of midmorning. One receives constant attention in the arctic when you are from without. Everyone wants to know if you are dressed properly. Are you warm? Do you have extra socks? How about water? Can I get you anything? Gloves? Mittens? Something for your face? These questions, of course, are generous and kind. Their intentions are good. They are also probing. Will this person be a liability? Will they endanger me? To announce one's foreignness in the arctic is to admit you are a danger to yourself and others, to summon the watchfulness of those in your company who would then decide whether you will be a burden or whether you might pull your weight. It's a quiet judgment passed without appeal.

An inordinate amount of time is then spent proving your mettle. This is the scrutiny I am met with this morning, as Jan and I trundle down a footpath to a large utility hangar. Inside are military trucks, snow machines (the word *snowmobile* announces my foreignness), mortar tubes, the casual Stingray missile. Two or three men in military fatigues scribble on clipboards, taking inventory of the mobile military stoves and cooktops used for field kitchens in winter. As they have since the invasion of Ukraine, the Swedish military is sending more resources to Ukrainian forces, whose winters resemble something out of the arctic. Much of what is being learned in Ukraine is standard operating procedure during cold-weather training for those preparing for war in the circumpolar north. Modern plastics are brittle in subzero temperatures, making newer hand-held radios unreliable; instead, use older equipment, and prefer wired over wireless communications.

Jan strides back to his car. We drive west, away from Kiruna. The drive is a montage of arctic forestry. Tall pillars of light in intervals along the road, hidden by trees, announce cars and their headlights in the distance. The churning of studded tires against asphalt. The slight diversion of the car as the wind pushes the trailer towing the snow machines from side to side across the roadway. The hulking moose standing along the road. Jan has lived in Kiruna all his life. He is a volunteer with the mountain rescue squad, a group funded by the local police, who see most activity around the start of each year, when the ski season begins in earnest, when the outsiders arrive.

At the base of Nikkaluokta, near the border with Norway, the temperature reaches minus 10 degrees Fahrenheit. Jan, all eyes, with the rest of his body wrapped and his face swathed in fleece, takes off, leading me into a whiteout. Soon we are blinded and stop the snow machines. A sheet of white separates us, punctured only by the glaring red taillights of the snow machines. Slowly, the snow begins to rise and the wind pushes drifts of white void against us. We move slowly, as through a cloud. Then the moment passes and overhead it is clear blue. We throttle up. Ptarmigans croon and dart out of the way. Down in the valley, the lakes are frozen stiff, like flat white puddles at the foot of the sharp ascents to the Västra Leden. Somewhere beyond, Áhkká, the largest massif in Sweden, segregates the passing tempest.

We speed across the lakes, the ice three feet thick. Soon Jan stops and drives a drill into the ice. Without warning and in quick succession, he gingerly clears one of the holes using a ladle to scoop out icy floaters. He grabs a few writhing maggots from a container, fixes them to a hook, tosses the fishing line into the water below, and jams his face into the hole. His parka hood covers his head and the hole, and for a moment I wonder if he can breathe.

There are many times one hopes to be alone in silence in the arctic, but this requires some planning and the powers of imagination. Great discomfort comes from the loud silences, the opaque expectations and desires, of two people who are yards apart and do not speak. After all, to be cold is simply another word for silence. Frigid behaviors are not unique to one place. These actions are human. The arctic is not so different from anywhere else, when viewed this way. For one to be welcome in the arctic, it is important to be proficient (at something, anything) and uncomplicated. Rather than saying you are grateful, you must prove it.

His head out from his hole, Jan pulls up an arctic char, then another, then some grayling. After two hours on the ice, we have a sizable catch. The sun wanes as we make our way back to Nikkaluokta. There is no path. There is no evidence of the two-foot-deep trail we blazed mere hours ago—the windblown snow has covered our tracks. It would be impossible for me to navigate without a guide, without someone who has been here before, who has lived in the north his entire life and has taken me along to teach me how to survive, to fish and to feel. My gratitude can be expressed only by not causing a hardship. By not flipping the snow machine, by dutifully changing wet clothes, by sharing the tobacco I carry, by keeping apace and being watchful of the wilderness around us. I am able to express my thanks, and in turn I learn much. The arctic is self-preservation personified, where collaboration saves lives, but where no one should be asked to needlessly risk their own life for the carelessness or indiscrimination of others.

By the time we reach the car and trailer at the base of the mountain, the snow machines' headlights punch weak cones through the early-afternoon darkness. The sun has long since set. We load the snow machines and undress as the car warms. Two tourists appear in a rental car. They roll down their window and ask where they might find moose. They ask if there are any here.

"Not here," Jan says. "Back on the road, we saw some earlier."

They ask what is behind us along the trails.

"Nothing," Jan says. In a sense, this is true. The mountain will not open for several weeks, when the tourism season begins and the mountain rescue squad is fully staffed. The couple glance around, say thank you, and return the way they came. After they are gone, I ask Jan why he did not tell them to go for a walk in the woods, along the lakes and trails, where surely they might see a moose or two. He removes the hood of his parka and wipes the ice from his goatee. "Because then I'd be the one sent to go find them."

OPERATION
TOUR DE HELSINKI

BORDERS ARE IMAGINARY. THEY do not conform to nature, though they often follow it. Human nature does not conform to nature. So a struggle ensues. The house I rent is about half-an-hour's drive or so southeast of Kirkenes, about 250 miles above the Arctic Circle, but soon I will head for the borderlands of three arctic nations. Nestled in a remote hamlet, the two-story home sits amid autumn foliage so bright it is an easy distraction for the unaccustomed driver, the white bark of the birchwood unfurling like scrolls on the fringes of the Scandinavian taiga. The house has had many iterations and hosted many guests over the years. Among those who rested here along the Paatsjoki River were Prince Philip of the United Kingdom; King Olav V, Crown Prince Harald, and his new wife, a "commoner," Sonja, of Norway; and untold numbers of legendary actors and singers. In the 1990s, it was a restaurant, around the same time the nearby Pasvik Nature Reserve along which the house sits was established by Russia in 1992, and by Norway in 1993.

The house is in the town of Skogfoss and is painted red, with a gazebo out front and one neighbor down a stone path. A nearby electrical substation vies for the airwaves against the babble of the Paatsjoki ("Pasvik" in Swedish, "Pasvikelva" in Norwegian, and so on), over which the home virtually hangs. Interior doors are adorned with scenes from Norwegian folklore, like a depiction of Trolltunga, a large rock formation said to be the remains of a troll who turned to stone in the sunlight, or of White Bear King Valemon, a handsome king bewitched into a bear by the princess who rides atop his back. She'd fallen in love and vowed to trick the troll witch into turning back time and giving him back his freedom. What lay between them was a curse, a dividing force that kept them apart and somehow always managed to bring them back together. Nations that have a deep connection to their arctic roots often possess parables apt for explaining today's changing North.

In a room crowned with mounted animal heads, photos on the walls show young women standing at a roaring riverbank, the whitecaps crashing against large rocks. And men with their boats. And felled birch floating along the river toward Melkefoss, a few miles north. The river at Holmfoss. At Herefoss. I quickly learn that *foss* means "waterfall," that these waters feed into the North. I also recognize, in the background of each photo, the Russian taiga, which I can also glimpse through the kitchen window.

A fork in the river is hemmed on either side by boundary markers (green and red for Russia, yellow for Norway) laid every twelve feet or so. The markers have in recent years become celebrities themselves, with social-media parody accounts of their own.

Increasing traffic from the Russian side—mothers and children on grocery runs, men in search of work, the occasional Russian war deserter, a purported spy or two—has led to greater scrutiny of this shared zone, primarily by NATO security planners. It was not to be forgotten that Russia was at war with the West: "That's for sure—we are saying that openly," the Russian representative to the United Nations says.

There are many reasons why this area, which includes a latticework of international arctic boundaries, is well guarded, highly divisive, and seen by many countries as geopolitically, if not outright militarily, strategic. In two separate exchanges with the Norwegian Ministry of Foreign Affairs, the phrase "the Nordic region poses no threat to anyone" is repeated to me as if it were a mantra slowly losing its meaning. Say it often enough and it may remain true, a stout fallacy. Yet a flurry of activity, from an increase in troops patrolling the borders along the Pasvik River and heightened scrutiny of those entering and departing through northern border crossings—like this one with Russia, where signs along quiet hillsides point to the city of Murmansk and the home of Russia's Northern Fleet and nuclear arsenal—say otherwise. "Our bilateral cooperation with Russia has been reduced to a minimum. We are neighbors, so we need to maintain some contact to minimize the risk of misunderstandings and unintentional escalation in our neighborhood," the minister of foreign affairs in Oslo tells me in an email. "We will continue to support people-to-people cooperation with Russia to the extent possible, but the possibilities are now very limited."

A PRISTINE, WINDING ROAD hugging the Pasvikelva eventually deteriorates into a bumpy byway, a series of gashes and divots in the pavement before transitioning into a gravel logging road two hours south of Skogfoss. Large yellow signs hung on fences warn of the territorial boundaries. After trekking across flooded plains and half-submerged wooden planks used as pathways, I find Oona in the shade of a makeshift lean-to, the worn pages of *1984* flayed open on her lap. She is in her late teens or early twenties and is dressed in the standard Norwegian border-guard uniform. Simon, also young and slender and likewise in his border-guard fatigues, places beside a smoldering firepit the military handbook he is reading and greets me. He stands at the sound of large wooden chimes clanking as I pass into the border zone.

"Have a good hike?" he asks. Old-growth Scots pine shade the area around their fire. The trees and bogs and rocks farther down the hill form a natural defense; nevertheless, a fence delineates where one country's national park ends and another's wasteland begins. Here sits the boundary line dividing east from west.

This area, a postage stamp of land comprising the northern borders and the three different time zones of Norway (GMT+1), Finland (GMT+2), and Russia (GMT+3), is marked by a seven-foot cairn, the *treriksrøysa*. The view afforded to Oona and Simon from atop Muotkavaara is desolate. Their demeanors—jovial and mirthful, like precocious children set free in a land of fairy tales—are at odds with the border they guard. Only this side is accessible to the public. Russia maintains deep and wide border zones, restricting access to military and police patrols. The nearby Kola Peninsula, because it is the home port of the Russian nuclear arsenal, is one of the most heavily guarded regions on earth. There are few people on the Russian or Finnish sides that live close to the border with Norway; soldiers like Oona and Simon stand in watch towers overlooking the scar of land—a clearing of open taiga where the river does not run—monitoring the Russian FSB agents who monitor them.

The border, this imaginary space demarcated by nations and their governments, has often shifted. It has been given different names, been overseen by different jurisdictions. For the vast majority of that history, stretching back a thousand years, the Sámi could herd their reindeer across the borders between principalities and regions without fear or retribution, living nomadically despite ongoing wars and territorial conflicts, a freedom more recently stripped from them. How confounding it must have seemed to those reindeer herders to witness fighting over something to which they were taught everyone had a right.

Borders have meant the most to newcomers. Humankind's long presence in the arctic serves as a reminder that the north was always seen as a frontier that must be claimed and conquered, explored and mythologized, home to tales like those of the princess and her bear, of who has the right to what and why. There was Pytheas, the ancient Greek mathematician and astronomer wrongly considered the world's first polar explorer, who in 326 BC helped mathematically equate where the Arctic Circle lies; Ohthere of Hålogaland, the Norseman, who rounded the northernmost point in Europe some time in the 900s; Sir Hugh Willoughby, in 1553, who sought the Northwest Passage; Henry Hudson, Willem Barentsz, Semyon Chelyuskin, Dmitry Laptev, Vasili Pronchishchev, and so on. Their stories include: ships trapped in ice; men and their complements starved to death; a woman's discovery of North America; maulings by polar bears; and violent clashes against welcoming Indigenous people. The last is a culture relegated to the background and punished by developments that would soon irreversibly molest and rob a region of perennial charity. What was to be made of such an uninhabitable, difficult region? How might the world benefit? The narratives eventually became tales of domination: blood-soaked sand, already black from countless volcanic eruptions, beneath whale carcasses exploited to help light the world, their blubber used for oil lamps; walrus tusks taken for ivory bric-a-brac; seal meat used for consumption and polar bear fur taken for the aristocracy—claims laid only to be challenged by the absence, or inaccuracies, of an explorer's logbook, like the ongoing debate over which explorer truly reached the North Pole first. It was the displacement by those explorers of Indigenous persons and children, hauled back to the smog-choked cities of Europe and North America to die, left for postmortems and scientific studies in various prep rooms.

For its part, Scandinavia's border history is long and shot through with fluctuation, instability, and intrusion. For over a millennium, the northernmost regions of Europe have seen constant territorial shifts, as Russia and Scandinavian nations fought to secure strategic land and vital resources. The Vikings of Sweden and Norway once ventured eastward into present-day Russia, forming trade routes and alliances that eventually gave way to conflicts, as the Slavic world solidified its power base. In medieval times, Novgorod and Swedish forces clashed repeatedly over control of Finland, as both sought to expand their spheres of influence. By the early modern era, the Kingdom of Sweden had emerged as a military powerhouse, waging several wars against Russia for dominance over the Baltic. The Great Northern War (1700–1721) marked a turning point, as Russia, under Peter the Great, decisively defeated Sweden and established control over much of Finland. The Treaty of Nystad in 1721 reshaped the borders once more, solidifying Russia's presence in the region and marking the decline of Sweden as a dominant force in northern Europe.

Throughout the nineteenth and twentieth centuries, the tensions persisted, and no treaty held forever, particularly in the far north. Norway and Finland, vulnerable to Russian expansionism, saw their territories encroached upon during the Russian Revolution and the Second World War, during which Kirkenes became a strategic focal point. The Cold War maintained the region's geopolitical volatility, with both sides heavily militarizing their northern borders.

Those defenses, patrols, and surveillances fail at times, on all sides, and intrusions are still a very real threat to the stable democracies across Scandinavia, all of which are today members of NATO. In January 2023, not long after Oona and Simon started their service, shots rang out across the border. A Russian military deserter named Andrey Medvedev had crossed the Paatsjoki into Norway. He had fought with Russia's Wagner paramilitary organization in Ukraine, and had escaped in what he recalled as a daring nighttime dash across the frozen river, the F.S.B. taking aim as he crossed time zones and borders. Norwegian authorities arrested him and discovered he had fled from the town of Nikel, thirty miles east of Kirkenes.

Nikel has itself changed hands thrice in the last eighty years. First, the Treaty of Tartu ceded part of the area to Finland from the Soviet Union in 1920. A British company was given exclusive mining rights and built the infrastructure and railway that would pull the nickel from the fells nearby in the 1930s. Then, during a preamble to the Second World War, part of the region was ceded back to the Soviet Union, which still allowed Finland to operate the mine. Nazi Germany, with which Finland was a co-belligerent without ever formally joining the Axis alliance, exclusively relied on Nikel for its nickel, which they used to harden the steel of their tanks and warships. Sweden fully ceded the land and mining rights to the Soviet Union when the Red Army occupied Nikel in 1944 and claimed the city of some 12,000 residents. Its sister city is still Kirkenes. Today, it is home to the third-largest nickel and second-largest platinum and palladium producer in the world. So it goes along the shifting boundaries and wavering prospects within the borderlands.

As we stand together looking out over Russia and Finland and Norway, Oona and Simon are quiet. It was a wild week to start their new jobs, and they were grateful for some peace; but episodes of excitement like that were not atypical, at least not in the preceding decade. Kirkenes had become something of a choke point for Russian campaigns of unruly antagonism across the high north. Take, for example, the refugees.

The cyclists arrived first in Kirkenes, in 2015. Blue and red bikes, no suspension, wide and tall handlebars with tires not made for snow. Astride their seats were migrants, their faces stretched into wide smiles, as they pedaled toward Norway.

Hundreds of refugees and migrants and asylum seekers from wars in the Middle East, those displaced by poverty and excessive heat leading to famine and drought across Africa, had made their way to Russia, and Moscow wanted them out. Before Kirkenes, they arrived in Russia, then they were sent by taxi to the dilapidated factory town of Nikel, visible from Oona and Simon's watchtower, before mounting the bikes for the half-kilometer ride to the Storskog border crossing. Neither the F.S.B. nor Norwegian officials allowed crossings on foot, so Russia handed them bicycles.

Russia's attempts to disrupt Scandinavian stability continue, as seen in campaigns around Kirkenes and Sør-Varanger, which were not isolated. They grew and became a continued recurring gray-zone tactic used by Russia to divert security resources and increase domestic pressure on undocumented immigration. The use of migrants as a weapon continued apace and increased the domestic pressure on undocumented immigrants in Norway and Finland.

"I'm afraid," one Scandinavian NATO commander tells me when discussing the region's instability. "Norway will not probably be the problem," he says. "It will be the border to Russia from Finland."

DOWN IN FINLAND, ACROSS its 830-mile border with Russia, Helsinki is overcoming several new waves of more than 1,300 migrants each (or nearly a third of the active-duty Finnish Border Guard), who were forced into the region by Russian authorities starting in 2023, so as to provoke disruption and test the limits of Scandinavian resilience.

Finland has some 50,500 bomb shelters scattered from the Baltic Sea in the south to the arctic taiga of the north. Five thousand of those shelters are in the capital, Helsinki, and can together house the entire city's population with plenty of room to spare—a veritable underground city in its own right. A majority of the shelters across the country are so robust that they can withstand nuclear strikes. Those would make sense for protection in conventional warfare, but they are not particularly helpful against the hybrid and asymmetric attacks spreading across the country from Russia, nor are they positioned to house short-term migrants.

It was for this reason, and a historic distrust of Russia—known for dishonoring its treaties—that Finland in 2022 formally sought accession to NATO and, in 2023, joined the defense treaty. "If you draw a circle of three hundred kilometers from Helsinki, inside that circle the vast majority of the Finnish population lives. Our national defense has mostly been oriented to defend the south," Colonel Matti Pitkäniitty, of the Finnish Border Guard (Rajavartiolaitos), tells me before arriving in the Finnish capital. (I once heard a Swedish soldier quip that Finland's entire defense plan was as simple as "Retreat to Helsinki!") That was then. "The value of the High North has grown exponentially after our accession to NATO," Colonel Pitkäniitty says. Their joining NATO brought its own problems, spurring Russia to increase its pressure on the country's less-guarded northern frontiers. "Those weapons in the Kola Peninsula, they were not aimed at Finland or Sweden. They are there for NATO and the U.S. But now, being part of the alliance, what's a problem for the alliance is a problem for us."

The Finnish Border Guard is a military organization, part of the nation's broader collective defense forces. During the Cold War, the organization's ethos was simple: Shoot anything on sight. Ever since the fall of the Berlin Wall and the dissolution of the Soviet Union, the force has rolled back its mandate and become something more like a traditional law-enforcement agency. "That was when people thought there wouldn't again be a war in Europe," Colonel Pitkäniitty tells me.

After Finland joined NATO, a Russian Foreign Ministry spokeswoman, Maria Zakharova, said that Finland's cooperation with NATO and the United States "will not leave unanswered the boosting of NATO military capabilities on our border." She added, "The responsibility for turning a zone of neighbourliness in the region into a zone of possible confrontation will fall entirely on the current Finnish authorities."

When I visit the border-guard headquarters, it is nearly May, and the barometer indicates snow. The storm leads to the death of a ten-year-old and cripples public transportation. Even the most innocuous, anticipated storms still bring surprises.

In the entrance hall at the old headquarters in downtown Helsinki, hidden along a side street crisscrossed by tram lines and electrical wires, and secured behind a metal detector, bulletproof glass, and a series of remotely locked windowless doors, a table ornament displays a miniature *treriksrøysa* set between two boundary markers (blue and white for Finland), a replica of the one I had stood beside with Oona and Simon. A blue-hulled model ice-breaker sits in a glass display case. On the walls are nine maps. Various hues of green and red and blue and yellow depict the historic border shifts of the European Arctic as it was segregated from, integrated with, then once more detached from the Russian Arctic.

Colonel Pitkäniitty's superior officer, Colonel Marko Saareks, a deputy chief of division at Rajavartiolaitos, settles into a seat in a quiet room on the third floor and reflects on Maria Zakharova's words—what have become more or less coined in the press as "Operation Tour de Helsinki," with hundreds of migrants being led by the F.S.B. to Finnish border crossings and leading the country to close its border with Russia. "We don't wait for it to happen," Colonel Saareks says. "We don't know with one hundred percent certainty whether it will escalate or not, but we have to prepare."

We are seated some 124 miles from St. Petersburg. He says they have tried the normal diplomatic channels to stymie the flow of cross-border threats (including the occasional military surveillance drone), but tensions never settle; they only seem to rise. Colonel Saareks describes a phone conversation with an F.S.B. officer who he says could only parrot Kremlin talking points. "They don't have any kind of free speech, so it's quite stupid when you know he's just repeating what he's told, ordered to say." The F.S.B. officers say they are doing their best; that they cannot prevent those in Russia without documentation from moving on to another country. Since then, the F.S.B. has outsourced border security to the Rusich Group, a far-right neo-Nazi paramilitary organization similar to Wagner, to "strengthen the border with Finland."

If the Russian authorities were so willing to let the others cross, why would Medvedev, the Wagner soldier who crossed into Norway, be the exception? Perhaps the migrants are better weaponized in these campaigns along the borders into Europe than along the front lines of the battlefields in Ukraine, to which Russia was recruiting felons, murderers, and North Korean soldiers. Or perhaps Medvedev is running a counterespionage campaign. He tries, unsuccessfully, to cross back into Russia and is arrested again, much like the tourists who visit the cairn and run circles around it for sport.

There are also borderlands in the sky. Finland has reported multiple airspace violations by Russian jets, heightening concerns about domestic security. Along the Baltic Sea, tensions seem de rigueur. Surveillance aircraft and patrols have increased, and the Baltic itself has become a zone of military focus, particularly when Finnish military officials announce they have spotted a Russian submarine near Helsinki. Since Finland and Sweden joined NATO, many officials and academics have referred to the Baltic Sea as "NATO lake," reflecting the alliance's increased presence and further angering Russia.

Russia is still not the only concern. China has been pushing for a larger role in the region, particularly through projects like a now-abandoned proposal to construct a tunnel running beneath the Baltic Sea between Finland and Estonia. Despite strained relations with Beijing, some Finnish developers tried moving forward with the project, seeing it as a crucial link for trade. Economically, Finland remains dependent on certain Russian imports, such as nickel and fertilizers, which fall outside wartime sanctions. Finnish officials emphasize the difficulties of cutting off this trade entirely, given its importance to key industries. The country's relationship with both Russia and China continues to be a balancing act, shaped, like the rest of Scandinavia, by security concerns weighed against economic realities.

Where the Finnish are placing their efforts, however, is in fortifying and renovating the citywide bomb shelters at a total cost of roughly $27.5 million, in anticipation of a wider arctic conflict spillover. But the more ambitious infrastructure project is the construction of a fence planned to run along a portion of the country's border with Russia. At six feet tall, it is expected to one day be equipped with barbed wire, loudspeakers, and night-vision cameras.

As Colonel Saareks leads me out through the hallway, I point to the colorful borderland maps. He nods, placing his hands at the square of his back. I ask if he thinks the borders will continue changing. "They have," he says solemnly. "And they can."

TO GET TO THE first 124 miles of fence, I pass the Venna Military Area, nestled amid stone escarpments. Though far south of the Arctic Circle, the spring light flays across farmland the way it does in the higher latitudes, the soft pale hue of dusk a reminder of the north's proximity. Then one passes Nuijamaa, the old turnoff for St. Petersburg. Military vehicles and trucks in convoys race down the highway. Farmers tend their fields and tractors.

So popular is the fence, constructed outside the city of Imatra, that the president of the European Commission, Ursula von der Leyen, visits the cowlick of land extending into Russia only a week before my arrival, during a celebration of its construction. The town still basks in the prestige and pomp of the visit.

Before reaching the fence and border area, Antero Lattu, the president of the Imatra City Council, wants to talk about the Russians and the fence. He will not join me at the fence, but says the main issues facing the city are "the refugees, because we have a lot of forest and thinly populated areas north from Imatra and also south. It's very easy to come; there isn't any kind of obstacle nowadays if you go north."

Now in his early seventies, Lattu remembers the days of the Soviet Union. There was a military buffer zone then, to separate the West from the East. "Putin now wants that back. And we in Finland decided we don't want to be in that buffer," he says, noting that Ukraine was supposed to be a buffer, too.

"But Russia decided they will take the whole country, that they must control the countries west from them." Imatra is a key exporter of pulp and consumer packaging, like cardboard. That's not exactly a key natural resource, but any disruption to the city impacts the economy, and with trade between Finland and Russia largely halted, Lattu says, Imatra is finding it difficult to manage its own small polycrisis against actual threats from just across the tree line.

This is easy to see. The road to the border checkpoint looks like it has already seen war: a supermarket sign is half hung, its Cyrillic letters dangling from electrical wires, and a parking lot where the buckling pavement plays host to blades of grass tearing through. A middle-aged woman walks a dog. A young mother pushes a stroller. A man walks with Nordic trekking poles along a path, his back to the border crossing. It is otherwise deserted. There is a light layer of snow from the recent storm.

My car is met with suspicion and weariness by two soldiers dressed in frumpy and ragged dark olive Border Guard uniforms, who moments earlier emerged from a ten-foot-long white shipping container surrounded by safety cones, warning signs, and various fences. They say the border is closed. I say I am there to meet with Captain Jyrki Karhunen, head of the Imatra Border Crossing Point Southeast Finland Border Guard District. The young men mumble my request into a toylike handheld radio. Arrayed around their belts are cans of riot spray, Tasers, batons, and black nine-millimeter handguns. They wear tactical vests. Several minutes pass. Captain Karhunen arrives in a blue sport utility vehicle, parks near the shipping container, and steps out. One of the guards grabs hold of one of the chain-link fence sections, lifts it from a concrete footing, and lazily salutes as Captain Karhunen steps through.

Captain Karhunen speaks in short sentences. He often defers to his subordinates, who shrug before giving the answers to a series of questions—answers which he then repeats. The real fence is still coming together, 2,600 feet farther along. The younger subordinate is still holding the section of fence, as though his captain will leave at any moment. I ask to see the fence.

Captain Karhunen says the border is closed. I say there was a press tour not that long ago. But it is not really finished, the fence, and the guard towers are not yet being built. Can I see it? No. It is closed. He says lots of people come, not knowing this.

As I prepare to leave, the wind picks up. A gust originating from the east blitzes the station. Leaves clatter along the pavement and the subordinate loses his grip on the fence; it teeters, then it collapses at our feet. I think of the myth and the princess atop her bear—how I interpret it as a story of devotion and commitment through hardship, and realize others will take away much more. Borders are like myths in that way; in today's world, they are open to interpretation, never fixed, and rather precarious places. The wind rustles the downed chain-link. Somewhere, a bird takes flight. A moment of silence precedes the glances shared among the four of us. We say nothing more.

THE TYPHOON

LIKE BORDERS, TYPHOONS EVOLVE. They are born, then they die and can be born again. The path each typhoon takes may reanimate the storm, giving it enormous strength and bringing about incredible changes at sea and on land. At midnight on September 11, 2022, Typhoon Merbok was born off the coast of Japan, deep in the Pacific Ocean. Where it would head, along the shores of the Bering Sea, from Siberia to Alaska, no one—not fishermen, not schoolchildren, not Navy SEALs—was prepared. Yet, through the serendipity of a hundred-year weather event, their lives would intertwine.

The storm coincided with a dramatic wake-up call for Alaska and U.S. Special Operations. Indigenous communities and myriad timber-framed homes along the Alaskan coastline were about to be severely damaged by a storm few anticipated. Special Operations Forces like the Navy SEALs had only recently begun training in the arctic, after decades spent in warmer climes fighting jihadist uprisings and Islamic fundamentalist movements in the Middle East. They, too, would be caught off guard. The Bering Sea, whose waters were warming and creating unpredictably strong weather systems, was in the thralls of extravagant changes. Aside from the SEALs and coastal villages, what lay in the typhoon's path was a $6 billion-a-year industry providing 60 percent of America's domestic seafood. The Bering Strait and its environs were experiencing a rise in commercial fishing; thanks to disappearing sea ice, vessels were able to stay at sea for increasingly long stretches. Those boats were joined by Chinese and Russian naval warships maneuvering off the Alaskan coast, welcoming new fishing vessels and more military patrols.

Dire warnings about the storm sent by email to residents in the Alaskan town of Nome by climate specialists went unheeded. Rick Thoman, a researcher at the International Arctic Research Center at the University of Alaska Fairbanks, warned of a superstorm with unprecedented coastal flooding. Thoman sent updates frequently, adjusting the models showing the strength and size of the likeliest storm surge flooding as new data came in. "It was clear five to six days away" that the storm would be severe, Thoman later tells me.

Merbok lived for roughly four days before transitioning into an ex-tropical storm somewhere south of Alaska's Aleutian Islands. The typhoon crossed into the Bering Sea on September 15. A day later, Merbok reached its lowest atmospheric pressure, creating gale-force winds, up to 90 miles per hour, extending for 250 miles in nearly every direction. Business owners in Nome, a village of 3,600 on the banks of the Bering Sea, secured propane and fuel tanks. Homes and storefronts were boarded up. Without adequate supplies to make sandbags, many residents used plastic sheets to seal doors and single-pane windows. Gravel was tossed along thresholds to act as a water barrier. Students went to school for a half-day; volleyball was canceled.

IN THE DAYS AFTER Merbok was born, there were footfalls and then a knock at their door. The men around Egvekinot, a port town and fishing village in Russia's Far East, knew the sounds well, found themselves practiced in the movements that came next. A quick glance through a window revealed a man and a woman dressed in green military uniforms. There was a slow retreat from the front door so as not to awaken the floorboards, to not alert the visitors: Members of Russia's internal-security agency and K.G.B. successor organization were out front.

POLAR WAR

Maksim Teiunaut—hazel eyes, heavy brows—and Sergei Nachaev—older, a bit taller and broader in the shoulders, with the same mirthful affect—were best friends from high school and had lived in the fishing village all their lives. Maksim had heard the knock at his door earlier that month, and on that occasion, too, he peeped outside and then quietly retreated. He was already convinced his home was not a place he wished to remain. "There was no choice," he tells me later. And the two set out planning their journey across the Bering Sea, in a thirteen-foot boat, to America, across one of the world's deadliest bodies of water. The war with Ukraine had reached their home in the far-flung stretches of Siberia. Their hometown of Egvekinot was literally shifting, under external stress from a military draft and the impacts of climate change. They could not stay, and they could never go home again.

Max, as his friends call him, was just shy of his fortieth birthday, meaning he fell within the age range for conscription into military service. Sergei, seven years older than Max, was in no real danger of conscription but faced legal trouble because of his vocal opposition to the government in Moscow. Max was making good money fishing. The average monthly wages were roughly $500. Max earned between $1,500 and $2,000, oftentimes taking his boat nearly 200 kilometers out from Egvekinot to find the best catch. Sergei, too, relied on fishing. But the good money was not enough to make them stay. After both men were visited by officers from the security services, Sergei called Max and they grabbed some necessities: a Garmin 276C and an Iridium 9575 Extreme satellite phone. They downloaded weather and map apps onto their phones, using the Windy.com app and Soviet military maps of the region. They expected there would be patrols at sea and along the shorelines.

Conscription may have seemed enticing—the pay was roughly $4,000 a month and the promise of land back in the country's Far East—if not for the near-certain death that awaited Russians and Ukrainians in the bloodiest shooting war in Europe since the Second World War. Russia had also begun offering evermore financial incentives to those who fought and survived, all in the hope of not having to issue draft orders like those Max and Sergei sought to avoid. "Free Lands in the North" was an old story. The government in Moscow had tried this before, in 2016, but with little development and no industry, no communications and no roads, there was also no appeal in gaining Siberian land at the cost of blood and limbs. And besides, Max and Sergei already lived there and saw how their prospects in such a neglected, weather-beaten region did not a future make. Those who lived through Russia's conscription rarely reaped the benefits. The only money returned to the Russian Arctic was in coffins, as part of the benefits paid to widows and the orphans of those killed in battle.

Maksim and Sergei pulled their four-meter-long *Sever* with "a good motor" out into the water.

TYPHOON MERBOK AND ITS destructive winds had concerned Alaskan communities along the Bering Sea since the National Oceanic and Atmospheric Administration (NOAA), whose own future was marked by uncertainty as Trump phlebotomized the agency, announced that it had formed in mid-September, so they, like the residents in Nome, shuttered and boarded up what they could. But these warnings did not deter a team of U.S. Special Operations Forces. Late on September 10, 2023, a UH-60 Black Hawk and a CH-47 Chinook arrived in Nome. The helicopters were shuttling Naval Special Warfare teams to and from St. Lawrence Island for a training exercise called Noble Defender. The command team knew the helicopters, piloted by members from the Alaska Army National Guard, could not fly in sustained winds of more than 45 miles per hour, despite predicted gusts up to 60 miles per hour. They took their chances anyway, and the SEAL teams parachuted onto St. Lawrence Island as Merbok approached. Children and families gathered to watch as the parachutists alighted onto their beaches.

In a prescient report from the Center for Arctic Security and Resilience published one month before the SEALs landed on St. Lawrence Island, researchers had warned of a need to increase Special Operations proficiency in the arctic regions. There were many recommendations. One found an increased need for Special Operations Forces (SOF) to not only train regularly in the arctic but also to base themselves there.

The American military segregates its forces. Many units are regionally focused—some to the Pacific, others to the Middle East, and fewer still to the arctic. Military units based in one part of the nation or world may get training elsewhere, but mostly they remain proficient in their one area and nowhere else.

Only in 2021 had Special Forces soldiers truly returned to the arctic to train for cold-weather warfare, a Pentagon official told me. The 2022 exercise was the second in a series of trainings that went from once annually to more than three times annually, and had brought Special Operations Forces to Alaska more frequently. "For a very long period the arctic was not a priority for special operations forces," a Department of Defense official involved in the exercise told me. "We're slowly shifting back toward there, but we still have a lot of capability—and a lot of lessons—to learn." Special Operations Forces, the official said, needed "longer periods of time in austere environments . . . not just to survive but to thrive."

Another recommendation in the report predicted that "solutions that facilitate domain awareness and improve climate and weather predictability . . . would be valuable for operational conduct, including mobilization and deployment." The report suggested, for instance, that "sensor placements can assist SOF decision makers and planners with critical data and intelligence to best advise geographic combatant commanders on SOF employment, especially in the defense of the North American Arctic homeland."

For years NOAA had sought to establish more sensors along Alaskan shorelines, which would help collect data on sea levels and act as early-warning indicators for typhoons and their sea surges. Of the more than 100 water-level sensors NOAA said were needed to predict and understand storms like Merbok, only two had been installed along 1,000 miles of coastline around Nome. The Bering seafloor itself has remained largely unknown to researchers, a mysterious blue expanse only 34 percent of which has been mapped.

Special Operations forces, the report stated, "must be located in the Arctic." One of the report's final conclusions was that soldiers and missions in the north required non–Special Operations forces support. One of the reasons behind the increased number of exercises was to gain familiarity with the territory, but also to make connections with the Alaska Native communities and partner with forces like the Alaska National Guard, the Alaska State Troopers, and the U.S. Coast Guard. This had been a cornerstone of Special Operations warfare since at least 1979, when 3,000 militant students stormed the U.S. Embassy in Tehran, captured sixty-three Americans, and held them hostage. Collaboration with local populations eventually saved those hostages, and it was a lesson the Special Operations forces remembered. It was something they learned again in returning to the arctic on the day that Merbok made landfall. "The expectation was much less than what actually happened," the Department of Defense official told me, despite Thoman's warnings.

PREPARATIONS—BY THE COMMUNITY and the Special Forces—were not enough. Warmer than usual waters had created a beast of a storm—nothing Alaska had seen in more than fifty years. When it hit the state and the wider region on September 17, seas surged by ten feet, washing away roads, lifting and displacing homes from their foundations, causing power outages, and substantial coastal erosion among forty Alaska Native communities along more than 1,300 miles of coastline. There were no reports of deaths, injuries, or missing people. The SEAL team encampment on the northeastern cape of St. Lawrence Island was a series of tents erected on the beach fortified with sandbags. They were torn away like tissue.

The waves from the storm slashed and battered the beachhead established by the SEALs, forcing them to find shelter inside three fishing cabins owned by Raynard Toolie, who lived in the community of Savoonga on St. Lawrence. It was the beginning of the exercise, and when the typhoon hit, the teams were forced simply to wait out the storm.

The purpose of the exercise was to regain lost competencies in the high north and to project American military might in the north, across Alaska. That part of the exercise was successful and became a lesson from which to learn. After the storm passed, the SEAL teams and other Special Forces continued training with Combat Rubber Raiding Crafts (CRRCs), known as Zodiacs, the small inflatable boats they used to circumnavigate the islands. The teams linked up with the U.S. Coast Guard Cutter *Stratton*, a national security cutter that had been patrolling the Bering Sea as part of the exercise. Aboard the coast guard vessel was another contingent of SEALs working to increase the interoperability and shared capabilities of both military units. But the security cutter was not designed to host Zodiacs, making the embarkation and disembarkation of the SEAL teams an exercise in and of itself—another lesson that could be learned, but one that would require redesigned ships and a budget to match.

As the *Stratton* and SEALs, along with about 100 other American Special Forces units departed from Alaska, Toolie watched as Maksim and Sergei pulled their boat ashore, nearest where his cabins sat on the beach at Gambell. Somehow, the two refugees had survived the 255-mile traversal of the Bering Sea during a typhoon.

Toolie, standing on the beach as the men disembarked from their weather-beaten boat, welcomed them as he did the Special Forces operators. He helped them unload their boat, canisters of gasoline, and a small rubber inflatable boat—similar to the one the SEALs used—which Max donated to the community along with a fishing rod. Toolie called the Department of Homeland Security and other local officials, and brought the two men, who were now seeking asylum, to the local police station.

Despite the carnage wrought on the community by the storm, the Alaska Native populations assisted not only American troops but also Russians. Through the cleanup of their own wrecked villages, they helped outsiders, a persevering thread of humility found across the arctic. Still, the arctic was no longer peaceful or the province of only those who lived there. It was opening up to all, and everyone was learning anew that this place of austere quiet was one of disruption.

"There's no going back," Max recalled telling Sergei during their crossing. "There's no extradition from the Don," he shouted above the waves, referencing a memorandum of Don Cossacks in the eighteenth century and a popular-culture reference in Russia. "There's no going back!"

A PERFECT STORM

A FILM OF ICE forms along the dock at the Coast Guard Air Station Kodiak, in Alaska, hemming into port the solitary 418-foot national security cutter. The USCGC *Stratton*, the same vessel that assisted the Navy SEALs during their training exercise on and around St. Lawrence Island, is homeported in Alameda, California. It is not designed for ice. I mill about with a gaggle of junior officers, who peer curiously over the weather decks down at the ice and water below. Soon the sun will rise and the ice will surely melt. After all, it is not that thick. Overnight temperatures have been hovering near zero this January and rising to just above freezing during the day. This is, after all, winter in Alaska.

The *Stratton* is preparing for a return to Alaskan waters, a three-month patrol along Alaskan shores in the Bering Sea. Its crew's mission, in part to offer assistance to maritime vessels in distress and to ensure fishing boats are in compliance with maritime laws, is also to patrol the U.S.–Russian maritime boundary line. The crew finds themselves steeped in an increasingly competitive environment. America's withering exceptionalism on the global stage, seen in its botched withdrawal from Afghanistan in 2021, its timid response to Russia in Ukraine, and the recurrent snafus of a greenhorn administration in the White House, was also unraveling across the arctic. America's arctic policy was built on the expectation that things would not change; so the nation divested, most especially at sea.

A PERFECT STORM

The U.S. Navy and its Special Operations forces frequently coordinate with the U.S. Coast Guard on missions as varied as freedom-of-navigation tours and drug-interdiction operations across the world's seas. In 2018, the navy decided to reevaluate its Arctic Strategy, which had already received a much-needed update in 2014. The U.S. Navy secretary at the time, Richard V. Spencer, told the Senate Armed Services Committee that "we need to have a presence up there." While Russia was busy building deepwater ports and, nearby, paving 12,000-foot runways for TU-95 bombers, the navy possessed no ice-hardened ships for similar missions or support. Aside from a secret submarine research facility in southeast Alaska and an annual submarine event, the navy maintains a visibly minimal presence in arctic waters. When asked why the navy was seeking to revise its strategy only four years after the release of its previous U.S. Navy Arctic Roadmap, Admiral John M. Richardson, then the chief of naval operations, said, "the arctic triggered it." In a point of clarification, Secretary Spencer added, "the damn thing melted."

Yet a decade after that assessment and dozens of reports about the need for an increased maritime presence in the arctic, only one vessel is tasked with Bering Sea search and rescue at any given time. The *Stratton* will be in arctic waters to enforce laws and treaties that have held the region together peacefully for more than thirty years—the very mechanisms that are endangered and tested daily by Russia and China, which send various ships and aircraft over these waters. The sea is large, and as commercial traffic is set to increase, the navy's ability to execute its mandates is stretched thin, but is backed by an increase in regional awareness and intelligence.

The U.S. Coast Guard, like the U.S. Navy, revised and released its own Arctic Strategic Outlook Implementation Plan in October 2023, based on a strategic outlook published in 2019. During this mission in the Bering Sea in January and February 2024, it was plain that the coast guard's best-laid plans were also evaporating. It had intended, among other things, to expand arctic surface capabilities and support arctic infrastructure by increasing its ice-capable cutter fleet; expand arctic communications capabilities by acquiring satellite connectivity for its regional operations in the far north;

and enhance its own culture of arctic innovation by creating yet another "arctic research plan" while collaborating with academic researchers. The embarkation in the Bering Sea would underscore the need for these changes, and much more. Most of all, it would reveal how distant the reality outlined in the implementation plan actually is. The struggles the U.S. Coast Guard faces in the arctic are the same ones they face globally, threatening national security in the Bering Sea, the Arctic Ocean, and waters beyond.

While a visit aboard the USCGC *Healy* offered a view of the geopolitical and practical hurdles remaining in international waters across the Arctic Ocean, a tour aboard the *Stratton* revealed a microcosm of America's degradation in its own arctic territory—a voyage marked by delayed responses, strategic miscalibrations, and the need to bridge the gap between the stated ambitions of the implementation plan and the coast guard's actual capability and capacity. Many of the ambitions laid out in the coast guard's strategy fail owing to insufficient funding and financial mismanagement. Low recruitment is also to blame. Many other nations' coast guards were also tasked with carrying out re-envisioned arctic strategies. But the United States tasked its own coast guard with a herculean effort, without the necessary support. The White House sought to build arctic infrastructure without a budget, to work with allies with whom it rarely collaborates, and to pursue climate science while excommunicating Russia. (Meanwhile, Finland, a NATO member, is building nuclear icebreakers for Russia.) The United States' late revitalization in arctic interest, after years of relative inattention and fifteen years of aimless strategies, contributes to a narrative of playing catch-up in a region where it once held significant sway.

BEFORE EVEN REACHING ALASKA, before the mission from Kodiak begins, the *Stratton* and its crew struggled. A metal cover protecting a P-100 portable diesel pump "exploded," one officer later tells me, after a violent wave crashed onto the fo'c'sle, or forecastle, the front end of the ship. "For me, it was another reminder to never become over-confident," a junior officer aboard the ship, who was tasked with securing the cover, wrote in an email home. "The ocean is a beat of it's [*sic*] own and it is important to remain ever vigilant and respectful." Then a series of mishaps plagued the crew as it arrived in Kodiak. A generator went out of commission, leaving the ship with no backup. The parts could not be sourced in Kodiak and would have to be flown from Anchorage at some point, while the ship was at sea. Small but mission-critical systems were breaking: a new fire alarm system repeatedly went off when there was no fire, sending crew members dashing to extinguish nonexistent flames at odd hours. A crane line snapped. The ship was listing constantly to its starboard beam. Worst of all, the one luxury in the officers' wardroom, the ship's only espresso machine, was out of order.

Adding to the past and potential woes of the mission—to the pressure placed on the one crew tasked with patrolling the whole of the Bering Sea, which grows as winter sea ice retreats farther each new year—the crew was prepared to encounter Chinese or Russian naval or paramilitary vessels. These frequent the area—America's Exclusive Economic Zone (EEZ)—with impunity. Or, they would detect the Russian long-range bombers and fighter jets that enter Alaskan airspace ever more frequently. The *Stratton*'s commanding officer, Captain Brian Krautler, and his crew would soon enter the third-largest marginal sea in the world, home to the deadliest commercial fishing industry. As if all this were not enough, as we ready to leave port, we learn that a typhoon is heading to where we planned to be.

IN KODIAK, BEFORE WE steam out to the Bering Sea, the days are filled with joviality, bowling and pizza, and jaunts to the local Fred Meyer grocery store. Some junior officers and non-rates (the sailors in training for a rating or apprenticeship) buy food and sundries; others buy a sled to ride at a nearby skate park blanketed in snow. Some junior officers attend classes, where they learn to identify regional fish and the gear used to catch them—trainings designed to help them while conducting fishery boardings and inspections in unfamiliar seas. They travel too frequently to develop expertise in the specific ocean ecology of any given region, though. The North Pacific Regional Fisheries Training Center is not far from the bowling alley on base. "Be safe in the War on Fish," an officer barks to a group heading to class one morning.

Lieutenant Peter J. Purcell leads the school, a warren of classrooms and experiential laboratories where coast guard personnel can get hands-on experience with the kinds of equipment—like a flow scale machine—they might find aboard boats that catch and process fish. They learn to recognize when the machines, which electronically weigh fish brought aboard a ship, are intentionally miscalibrated to allow the fishing vessel a larger than legal catch under maritime law. There are also mock control rooms to simulate what a boarding officer might encounter on a ship's bridge. They practice reporting back the catch of the vessel and tallying any infractions to pass on to regulators at the local Coast Guard District and NOAA.

The mock control rooms do not look like those that can be found aboard a modern fishing vessel, though. The equipment was installed in the 1990s and is vastly dated. "Everything in there is just for show," one of the ten instructors at the school tells me. "None of it works," Lieutenant Purcell adds. "There's a lot of realism we'd like to bring to the table but can't." The school simply does not have the funds, meaning law-enforcement officers often encounter equipment they have never seen before.

Prior to leading the school, Lieutenant Purcell had deployed with the Marine Forces Special Operations Command in the Indo-Pacific. He knew that the Polar Security Cutter program, for which the military was planning to build several new ice-capable security cutters—the next-generation of arctic variant of the *Stratton*—was consuming the lion's share of the U.S. Coast Guard budget. Coast guard officials have repeatedly testified before Congress that the service requires at least six polar icebreakers, three of which would be as ice-capable as the *Healy*, which has been in service for more than three decades and is frequently under repair. The first boat under the Polar Security Cutter program was supposed to be delivered by 2024. The new estimated arrival date, coast guard officials tell me, is more likely 2030. (President Trump's "One Big Beautiful Bill" of federal spending allocated funds for a proposed seventeen icebreakers, twenty-one cutters, and more than forty-five patrol aircraft, which coast guard officials said could take fifteen years before the first ships were even commissioned. It was not clear whether this was in addition to the already-delayed Polar Security Cutter program or simply to revive it.) "Once we have the detailed design, it will be several years—three plus—to get completion on that ship," Admiral Linda L. Fagan, then the commandant of the U.S. Coast Guard, told Congress in April 2023.

The Government Accountability Office (GAO) has for years warned that the U.S. government, including the U.S. Coast Guard, has made serious miscalculations in its efforts to increase U.S. domain awareness and capabilities in the arctic. For one, the coast guard's acquisition process for new boats is hampered by continual changes to design and a failure to contract competent shipbuilders. Moreover, the GAO found that discontinuity among arctic leadership in the State Department and a failure by the U.S. Coast Guard to improve its distribution of assets and funds "hinder [the] implementation of U.S. Arctic priorities outlined in the 2022 strategy."

Lieutenant Purcell appreciates that the coast guard's mandate extends beyond fisheries management. "Food stock is an important part, but is part of a much bigger picture—protecting our national security interests," Purcell says, noting that things like transshipments, where fish and commerce are illegally transferred from one country to another at sea, are on the rise in arctic waters, as elsewhere. Oftentimes, with so few assets, only about 10 percent of illegal activity is ever caught. "Force projection is an important part of military diplomacy. Our force projection in the arctic is garbage," he says. I ask, "Is there support among legislators and lawmakers to adapt to the changing arctic?"

"It's not all gone. But what are we going to do . . . wait? We'll be speaking Chinese by then."

ACROSS THE BASE FROM the fish school, a group of six junior officers make their way to Hangar 1, off the airstrip. They are embarking on an hours-long flight operation to glimpse a sliver of Alaska's impenetrable remoteness, across which the coast guard is tasked with providing search-and-rescue support. On this day, with clear blue skies and mild wind, the flight path will traverse the Alaska Maritime National Wildlife Refuge, north to the Lake Clark National Park and Preserve, west of Anchorage.

To put the vastness of the region, and indeed the breadth of the coast guard's mission here, into scope and grasp its complexity, consider this: There is enough land here for each of the state's 733,000 residents to occupy their own square mile. If you were to stretch out your right arm, thumb extended down, index finger outstretched, and the rest of your fingers curled inward, you'd have a portable map of Alaska. Kodiak is located in the crook between the thumb and index finger. The index finger would represent the Aleutian Islands chain and the downward thumb would be Southeast Alaska, including the capital, Juneau. From Kodiak to Kaktovik (the top of the hand representing the North Slope and Arctic Ocean shoreline), near the border with Canada, is the air station's coverage area. In total, that area is more than twice the size of Texas.

Inside the hangar, three white-and-red HC-130Js with two massive black propellers on each wing glisten as though recently painted. Before it became a linchpin of the coast guard's operations in the north, the Naval Air Station was the coast guard's first permanent presence in the region, commissioned on June 15, 1941. On April 17, 1947, the first detachments patrolled here. The six junior officers from the *Stratton* take their seats in a small anteroom off the hangar as they wait for instructions. On the wall nearest several rows of orderly seats, a plaque bears a quote from bush pilot Beryl Markham's *West with the Night*: "But for a little while, this is the place for us—a good place, too—a place of good omen, a place of beginning things—and of ending things I never thought would end." Kodiak Air Station is the introduction to the arctic for many in the coast guard. It is oftentimes also their last posting to the high north, and with them goes the knowledge of the region and its environs gained during high-latitude operations.

Some, like Lieutenant Caitlyn Gever, a MH-60 Jayhawk helicopter pilot with a blushing of freckles, have been here longer. She'd conducted two missions the night before our flight, one a water rescue and the other a medivac. During a wilderness first-responder course I attended in Iceland, an instructor told our class of glacier and arctic tour guides that "seconds never count and minutes rarely matter," but Lieutenant Gever says, "the minutes matter when it's cold." Even with a speedy hour-to-wheels-up response, time and distance are both challenges. The flight time from Kodiak to Utqiagvik, the northernmost town in Alaska, is some nine hours and requires a stop for refueling. "We go as fast as we can, but the response is slow," Lieutenant Gever says.

Time is nevertheless abundant in the north. The six junior officers from the *Stratton* sit and prod their phones as they wait to take to the skies. So green were some, so new to even the coast guard, let alone the cold north, that they wonder aloud what kinds of missions the HC-130Js run in Alaska. They knew search and rescue—but what else?

"They also drop care packages out the back," one junior officer says.

"Did you learn that from *The Guardian*?" another junior officer asks, referring to the 2006 movie about coast guard rescue divers starring Kevin Costner and Ashton Kutcher.

"*Call of Duty.*"

A crew member in a drab green flight suit enters the room and ushers the group to a waiting plane. The *Stratton* crew members are bundled in layers of military-issue clothing, though I cannot stop myself from pointing out they are wearing Helly Hansen jackets for their three-month patrol in Alaska. Even Norwegian apparel in the arctic trumps American alternatives.

The HC-130J breezes over Shelikof Strait to 16,000 feet. In the cargo hold, next to a bucket surrounded by a shower curtain that serves as a bathroom, the junior officers line the walls in seats of red webbing. Some listen to music with their earbuds, others have ear plugs, or "gooies," and some play games on their phones. Two are asleep. One sips on an e-cigarette.

A blur of white-capped mountains, patches of fog, serpentine rivers, and woodlands stretch toward the horizon. Practice air drops are canceled for the day, but the rear tailgate opens and a rush of cold air whips through the cargo hold. The engines roar louder. As we descend along the western edge of Lake Clark, the plane's infrared nosecone camera focuses on a plane landing at the Port Alsworth airport. The HC-130Js were recently upgraded to enhance communication between Department of Defense and Homeland Security operations across the region, which includes parts of the Pacific Ocean. We level off at 2,000 feet over the lake.

Four of the junior officers strap into a gunner's belt running the width of the cargo hold. One dangles his feet over the tailgate. Two others follow. The plane glides over the lake, then banks left into the valley surrounding the Tlikakila River. On either side the sky recedes, replaced by towering rock facades. As the plane nears Lake Clark Pass, with Double Glacier under the right wing, the pilot ascends sharply. Two of the plane's crew members, connected to the cockpit by coiled lines attached to their headsets, tilt like gyroscopes and recenter themselves. The mountains shrink as we rise.

THE *STRATTON* HAS A near-full crew for the coming patrol. Many will learn for the first time what it takes to operate in Alaska, overcoming frozen helicopter tie-down hooks and fishery boardings on vessels covered in ice. When their shore leave, or "liberty," ends, the *Stratton* crew is back aboard the ship the night before departure.

Life aboard a ship is one of relentless noise: a clock ticks, a radiator hums, announcements crackle, hatches squeal open and slam shut. Metallic wheels grind on worn steel; heavy footfalls echo through narrow corridors and ladder wells, as does the clatter of plastic dinner trays dumped into the scullery. The soft gurgle of pipes sluicing various liquids at various viscosities—diesel, wastewater, liquid gases of all kinds—from forecastle to bridge to cabin along the gunwales to the stern, a 475-foot trip from tip to tip. Some of these sounds are knowable, identifiable. Others are the inexplicable howls, whoops, whirls, flushings, and whiz-bang-pops of mechanics and seafarers alike.

A PERFECT STORM

The typhoon is approaching and Captain Krautler is eager to get under-way. He is perhaps a choice commanding officer for this winter's patrol, given his experience aboard buoy tenders and other coast guard vessels off the Alas-kan coast. The ship recently returned from the Indo-Pacific; now its sights are set on the Aleutian Islands, considered at least by one standard to be part of the arctic, though it is many degrees below the Arctic Circle proper.

Throughout each day, announcements are made through the pipe.

"Now. The evening meal is being served. All hands."

"Now. First call, first call. Colors detail. Man your stations."

"Now. On deck, attention to colors."

I climb to the flight deck and watch alongside three junior officers who salute as the ship's star-spangled banner sinks.

"Now. Carry on. Carry on."

And with the announcement went the day's light.

Later, at a pre-departure operations briefing in the windowless ward-room, there is frustration about a lack of supplies. "It was a little stunning Kodiak had only one gallon of antifreeze," Lieutenant Commander Jason Helsabeck, the *Stratton*'s operations officer, says.

Section chiefs brief Captain Krautler on the plans for the coming days: continuing flight drills, small-boat drills, eventually some fishery inspec-tions. Captain Krautler gets an update on the engine, the fire alarm system, the crane maintenance, and the timeline for when parts for the generator will arrive. Then the discussion turns to weather.

"Weather is king because safety is king," Captain Krautler says. "Starting Saturday, it's going to get ugly. Then at that point, we're in it." To a gathering of the crew later, the captain says confidently and with a bit of excitement, "It's going to get real nautical."

But something unhinges his boisterous posture. I notice it in the first few days, and I cannot quite explain my feeling that the captain's intention to ride out the storm by "hiding with pride" is both practical and fueled by a deeper fear I do not yet understand. The next morning, as we ready to depart, fear-ful of the translucent layer of ice forming around the boat, Captain Krautler approves the use of a local tugboat, at the expense of tens of thousands of precautionary dollars, to weave in front of the cutter, breaking up the wafer-thin shards of ice. We see those ice shards from the weather decks as the *Stratton* steams away from shore, the ominous hum of its three engines churning a wake, starboard from port at 10 knots and into the Gulf of Alaska, westward toward the North Pacific, then straight into the Bering Sea.

ON OUR FIRST DAY at sea, a fuel leak in engine 2 shuts the *Stratton* down. A $4 bolt is broken. Not a general emergency, but the ship grinds to a halt and a nonregulation bolt is all that is on hand to fix it. Meanwhile, on the flight deck above, the crew is practicing touch-and-go landings. An MH-60 Jayhawk arrives, lands, then departs. Lieutenant Boatswain's Mate Christopher Karpf (BOSN4) ginned up a slurry of calcium magnesium acetate to help deice the patrol boats and the helicopter pad tie-downs, called padeyes. When the engine is repaired, the law enforcement officers practice fishing-boat boardings on the *Stratton*'s stern. A morality officer places "happy lights" around the ship to help the crew overcome some of the shorter days and encroaching darkness. A week passes and the seas are mostly calm.

In the wardroom, the television is tuned to the Armed Forces Network, with its rotating commercials about enlistment and pride and how anyone, even the chefs (known in the coast guard as "culinary specialists"), might discover acts of terrorism anywhere, urging "constant vigilance." Up one deck, in the Combat Information Center, the screens are marked "Secret" and some are purposely dimmed as I stand behind them. Blue and purple lights illuminate the room. Captain Krautler sits in an elevated chair in the center. "Who's who in the zoo?" someone asks inside the darkened space, and the officers set about planning the next week of fishery inspections, collaborating with NOAA and other agencies to identify which boats are at sea, which of those have past violations, and which have not been boarded in a while.

Chicken cordon bleu is served for dinner. The culinary specialists call them "hamsters." The ship steams slowly toward Unimak Pass, 226 nautical miles away. The narrow strait is some nine nautical miles wide and part of the Alaska Maritime Wildlife Refuge. It is undergoing and anticipating the same flurry of activity as the rest of the arctic. A thawing and melting northern ice pack will bring more activity to these waters. Every hour of every day, commercial vessels traverse the pass, the shortest shipping route from the Pacific Northwest and western Canada to China, Japan, and South Korea. Maritime shipping activity has increased more than threefold since 2008, when roughly 100 vessels transited the Bering Strait, entering the swath of water south of the North Pole, east to Banks Island in the Canadian Arctic, and west to the New Siberian Islands in Russia. Alice Rogoff, publisher of *Alaska Dispatch* and then the wife of billionaire David Rubenstein, told President Barack Obama in 2015 that this region, thanks to climate change, would be destined to become the next Panama Canal. "Most at the Academy said China or Russia," when asked about the greatest threats to national security, Lieutenant Matthew Kenkel tells me one day as we watch touch-and-go helicopter landings from the flight hangar. "The smartest kids in class answered, 'Climate change.'"

According to the various dots and blips on the radar screens indicating ships at sea and nearby ports, those visions seem to be bearing out. More vessels are exploiting the declining ice across the Bering Sea. The *Stratton* remains as a transient in these waters, something the Chinese, Russian, and local fishing boats recognize when they see its white hull and U.S. Coast Guard–red striping. It is the outsider. "They know we don't reside up here. They know it far better than us. We're not trying to stump the chump," one senior officer tells a room of law-enforcement boarding officers during a briefing. If they humble themselves, "inspections will go a lot further for you guys."

The lack of familiarity with the territory is a fairly modern development, or deterioration, in the nation's history of arctic operations. The first iteration of the U.S. Coast Guard—the Revenue Cutter Service, founded after a proposal by Treasury Secretary Alexander Hamilton in August 1790—long held sway in northern waters. After Alaska was purchased from Russia in 1867, revenue cutters frequented the Bering Sea. The cutter *Rush* performed a similar mission as the *Stratton*'s Bering Sea SAR Guard, called the Bering Sea Patrol in the early 1900s. The *Rush* was responsible for navigating the Pribilof Islands, which the *Stratton* will skirt during its patrol, to protect native seal herds living there. Seal poachers were often thwarted by the *Rush* and its crew. Illegal hunts were made more difficult by virtue of *Rush*'s presence. So successful was the *Rush* in these waters that its legacy lives on today in the phrase, "Get there early to avoid the *Rush*!"

AS WE MAKE OUR way through the corridors, the wind is expected to reach 58 miles per hour—enough to drench the sides of any boat and freeze that ocean spray into thick curtains of ice, like dead weight. The likelihood of multiple search-and-rescue operations is higher now, the risk to fishing vessels greater, but the brewing storm is "everywhere we want to be," the engineering officer says. The likelihood of search-and-rescue operations amid the growing squall means it may be more difficult for the *Stratton* to reach a vessel in distress, a worry Captain Krautler articulates but I do not understand. Is that not why we are here?

We pass Cold Bay, home to an auxiliary coast guard base and refueling station, and one of the longest airstrips in the American Arctic. There are many small landing strips along the Aleutian chain and greater Alaska. Most are simple gravel or dirt runways. Many more were built across the state at the start of the Cold War, but in 1998, the Pacific Air Forces returned most of them to the land and tribal leaders. They no longer serve their use, not because the Soviet Union collapsed, but because neglect left the technology and infrastructure on those bases and stations of little strategic value. Even along the Aleutian chain, bases became overgrown by weeds—dump sites for naval and air force waste.

An MH-60 carrying the generator parts departs from Cold Bay and delivers them to the ship. That helicopter crew travels on to Dutch Harbor, then returns to the *Stratton* to refuel. The fueling truck at Cold Bay "had a casualty," so the *Stratton* becomes the only nearby fueling station—a necessity for maintaining search-and-rescue mission readiness between Kodiak and Dutch Harbor. Even the infrastructure that is still used in the arctic is ailing.

A gust of wind carries a powdering of snow that turns to raging sideways snowfall, disrupted only by the helicopter's blades and the updraft.

"It looks like Alaska," Captain Krautler says during helicopter operations as he watches from a tower over the flight deck.

"This is my first Alaskan snowfall," Lieutenant Karpf says.

"It won't be your last."

Several of us are huddled into the small space, the helicopter blades looking as though they might tear through the glass windows separating us. The weather here can degrade quickly, and it does. Visibility disappears, at once evaporating into a heavy cloud of misty fog.

On one of these drills, after the deck crew runs out, ducking beneath the whirling blades of the helicopter to secure the chopper to the deck, the helo's side hatch slides open and a bumbling man with a handheld camera bobs across the flight deck into the flight hangar below.

"Oh, great," the *Stratton*'s engineering officer says. Several heads in the room bow in dismay.

The cameraman is with a film crew for the hit Discovery Channel melodrama *Deadliest Catch*. For nearly two weeks, we only run drills, never any law-enforcement or fishery boardings. With the brief arrival of the film crew, I start to glean why. There is much talk of the weather and the "storm glass" on the bridge. The instrument, which some claim helps predict the weather, is a clear glass filled with liquid. When the liquid clouds or crystallizes, it is a decent indicator of a gathering storm. And it clouds and clears as we approach and depart the typhoon's path. My feelings about Captain Krautler's apprehensions are confirmed when I return to my berth one night to find a copy of a self-published investigation into the death of a young coast guard sailor who died in the Bering Sea, on a similar security cutter, in similar conditions, not that long ago.

It was November 11, 2013, and the USCGC *Waesche*, also based in Alameda, spent its first patrol in Alaskan waters as part of Operation Arctic Shield, a largely summertime arctic exercise involving several branches of the military, among them the Navy SEALs and the U.S. Coast Guard. "Our annual operation in the Arctic provides us with an opportunity to appropriately address the threats and hazards in the region," Rear Admiral Thomas Ostebo, commander of the Coast Guard 17th District based in Juneau, said that year. "Our presence also allows us to listen to tribal and local government concerns and better understand what the emergence of the arctic means for the nation and the state of Alaska." They learned more than he anticipated.

The *Waesche* was similar in design to the *Stratton* and other national security cutters. At the stern of both ships there were two hydraulic sliding doors that opened onto a notch where a small boat, typically a covered Long Range Interceptor, is released and returned. The launching of the boat requires the doors to open and a deck-force sailor to pull a mechanical pin, releasing the boat backward into the cutter's wake. To return the boat to the cutter, the patrol boat motors up into the notch and is latched in by a capture line and net. While heading to Dutch Harbor, the *Waesche* was diverted to respond to a search-and-rescue mission involving the *Alaska Mist*, a mainstay on *Deadliest Catch*. The *Waesche* arrived at the scene around 4 p.m. Though the *Alaska Mist* had lost propulsion, a sister ship, the F/V *Pavlof*, was already on scene. "Command, control, and communications equipment" aboard the *Alaska Mist* were working, according to internal coast guard emails and a report by the coast guard's Major Incident Investigation Board. A helicopter crew from Cold Bay was launched and a commercial towing vessel, the *Resolve*, was delayed but en route.

The *Waesche* had been ordered to the scene to rescue nonessential personnel, and the commanding officer sent out a small-boat team to rescue the *Alaska Mist* crew. The sea state was difficult, with winds up to 51 miles per hour. Water crashed into the open fantail and notch. With the first group from the *Alaska Mist* onboard the small boat, it made its return for the notch. Back then, in order for the capture line to be seated properly in its receiving hook, or horn, on the bow of the interceptor, a sailor had to stand on the bow to guide the rope into place. On that day, the sailor was Boatswain's Mate 3 Travis R. Obendorf.

What happened next is the subject of incident reports and witness statements, and an official investigation. Obendorf was hurled twice into the net and slammed violently twice into the steering column of the small boat. He died from his injuries. The rescue, BM3 Obendorf's mother later wrote, felt like it was for the benefit of the *Deadliest Catch* and viewers watching at home, as she says the airlift rescue was delayed a day until the television crew, which was regularly aboard that ship, arrived. It was a disaster and a reminder that ultimate responsibility lies with the captain of a vessel, the decisions they make, and whether those decisions concern boarding instructions or requests for tugboats in advance of ice. It was also a reminder that arctic waters are no place for an inexperienced crew with an insufficient tool set and support for operating in the region.

The presence of the film crew appearing aboard the *Stratton* now, however briefly, the ominous warnings of the storm glass and weather monitors, and the callowness of the crew, raised the specter of mistakes made by those who came before Captain Krautler. After the death of Obendorf, modifications were made to the cutters and their notches; the hydraulic doors now close in seconds, not minutes. Nevertheless, on a glass window near his seat on the bridge, Captain Krautler writes in dry-erase marker, maybe as a reminder to himself, or to his crew, of the importance of "accountability." The verbiage and the note itself reminds me of the other notes he scrawled on the title page of Obendorf's mother's memoir. Weeks before this embarkation to the Bering Sea, two Navy SEALs drowned in the Arabian Sea off the coast of Somalia. I imagine everything seems to Captain Krautler like a tempestuous warning.

"THE LAST TIME *STRATTON* was able to conduct winter boardings in the Bering was in 2021!," an internal newsletter sent to members of the crew and their families declares a few days later. The message rings hollow. It seems that in the few years between then and now, that competency and comfort have atrophied, no matter the optimistic internal and external messaging. One day on the bridge, I approach Captain Krautler after an aborted fisheries inspection. He frets that it is too cold, the seas too dangerous and unfamiliar. But the seas are not so bad, the wind mostly calm. I tell him I'm prepared to board any vessel he deems fit, to see the crew in action. I tell him I have done it before, in worse conditions.

"My guys haven't," he says, his voice rising as we stand on the bridge. This was true. Many of his boarding officers are hard-pressed to locate even the pull cord on their inflatable life vests. As he turns to walk away, he doubles back and says, "We are an arctic nation that doesn't know how to be an arctic nation."

An announcement comes over the pipe. "Now. Set LE Phase Two."

There are several successful boardings in the days that follow, but many of the crews tasked with climbing the pilot's ladders onto various fishing vessels are frustrated. They are dismayed by the weather conditions, nervous about whether some of their team members are trained well enough to embark on the boardings. On one long boarding, which takes upward of six hours, a culinary specialist wants to send hot chocolate to the small-boat crew. She cannot. The *Stratton* is without thermoses. "I don't think we're all experienced as we should be for fisheries inspections in winter in Alaska," one of the senior boarding officers says during a briefing after the boarding. Another officer, Lieutenant Mike He, describes one boarding as "embarrassing."

The crew needs a break, is looking forward to the first port call and liberty since departing Kodiak. After another frustrating boarding, late one evening over mid-rations of burritos and tortilla chips dipped in peanut butter, I stand with two junior officers on the bridge for the night watch.

A phone rings and Lieutenant Karpf answers. There is silence. The bridge is dark, illuminated in spots by small red lights strategically placed around the bridge and the dim glow of a radar screen. Few people are awake aboard the *Stratton*. Then Karpf doubles over in laughter. He asks another lieutenant to take control of the ship as he goes down to Lieutenant Commander Helsabeck's room. The rest of us exchange curious glances, though it is hard to tell who is who in the dark, despite our close proximity. When Karpf returns, we get the full story. Returning to his room earlier, Lieutenant Commander Helsabeck had been worried when he saw a pair of boots sticking out from his bathroom. He was surprised, believing someone was using his toilet. That was not the case. He called for backup because someone had placed the flotation dummy in full dress on the commander's toilet and he requested help removing it.

THE TYPHOON NEVER MATERIALIZES for the crew. At least not in the path of the *Stratton*. In the waning days of the patrol, the crew darts back and forth across the Bering Sea—to catch a suspected poacher of a Steller sea lion, to write a ticket for a vessel that has no AIS transmitting its location, to chase a boat that is suspected of drug smuggling, and eventually to catch a criminal wanted in Idaho for assaulting a child. Though we never travel north of the fifty-fifth parallel, there are occasional fears of geopolitical clashes. At one point, a few days before reaching the first port call in Dutch Harbor, an unidentified vessel appears on the radar. The crew believes it to be one of China's long-range fishing vessels, part of its government-backed "blue-hull" civilian naval militia. It is instead an American fishing vessel. Then the degaussing system breaks, leading to leaky pipes across the ship.

What will not appear on reality television shows like *Deadliest Catch* is the increased value the United States has placed on the arctic assets it has— and wishes it had—without devoting the necessary attention or budgets to attaining them. A month before the *Stratton*'s deployment, the State Department released a set of geographic coordinates "defining the outer limits of the U.S. continental shelf in areas beyond 200 nautical miles from the coast." What is known as the extended continental shelf (ECS) stretches out beyond Alaska into the Arctic Ocean. Many countries have made similar claims under international law to help manage, conserve, and profit from northern latitudes long ignored. With summer sea ice diminishing annually, and predictive modeling showing the Arctic Ocean could see ice-free summers by 2050 or even 2030, there's a new urgency to stake a claim to the spoils of the rapidly melting arctic. The land rush for natural-resource domination up here is also at odds with what's in need of repair.

The continental-shelf claim stands as the first significant legal push to recognize a domestic arctic interest. And the U.S. Coast Guard is tasked with an outsized role in executing a new vision for the north, not only performing law enforcement but also ensuring the safety of boats and their crews, who provide the nation with a majority of its domestic fisheries, as well as the oil and natural gas infrastructure on federal lands that generated $4.202 billion in 2019.

Determining the ECS's outer limits requires data on the depth, shape, and geophysical characteristics of the seabed and subsoil. Both NOAA and the U.S. Geological Survey are responsible for collecting and analyzing the necessary data. Data collection began in 2003, and having yet to be completed, it constitutes the largest offshore mapping effort ever conducted by the United States. One of the U.S. Coast Guard's congressional mandates is to support projects like those defining the continental shelf—more work for the coast guard. One junior officer, referencing one of the sayings often touted by coasties, tells me, "I know we're supposed to do more with less, but it's hard." And still, without the resources, the U.S. Coast Guard's mandates only seem to grow.

An announcement comes over the pipe: "Now. There is a bald eagle perched on the bow of the *Stratton*."

When we reach Dutch Harbor after nearly two weeks at sea, the crew has learned only by doing. More storms are predicted, harsher winds at 80 miles per hour, and tsunamis. In Dutch Harbor, the importance—historic and contemporaneous—of Alaska and the larger arctic to the country's defense is on display. Concrete pillboxes line the roads and trenches dug during the Second World War mark the hillocks around the harbor, the nation's largest fishing port by volume for the past two decades. The Nazis had their Atlantic Wall, and the Allies had their Aleutian chain. The global arctic is not so different. This was just another pit stop for the *Stratton*. As for the future of Alaska, that falls beyond the scope of the U.S. Coast Guard's mandate.

Much of Unalaska, known by its contemporary name of Dutch Harbor, is struggling to develop. Weather-beaten buildings line the roads from the port into town. The mayor of Dutch Harbor tells me projects ranging from fish-processing plants to geothermal exploration have been hindered by agencies in Washington, which also decided to award Alaska's next deep-water port to Nome, rather than building it here, at the gateway to the American Arctic and the port of call for the lone patrol ship tasked with its security. A deep-water port already exists in Port Clarence, seventy miles north of Nome, though it is a naturally occurring port and has fallen into disrepair. But Nome is more tourist friendly, so that is where the money goes.

"I don't think they're prepared for what they're dealing with right now," Vincent M. Tutiakoff Sr., the mayor of Unalaska, says of the coast guard and of the government more broadly. Dutch Harbor is a port of refuge, where a ship in distress can restabilize and protect its cargo and crew without threat to the environment or marine traffic. Even a port of refuge is neglected. "They have chosen Nome to be their big port of the future. And that's fine, but in order to get to Nome, they have to come through here," he tells me. "They need a container facility, a storage facility, which we have here—we have two. Military is going to need a place to come into to get out of the ice."

Tutiakoff tells me a team representing Elon Musk had visited in 2022 to explore the possibility of geothermal engineering projects, but Musk's people no longer return Tutiakoff's calls. I ask how much the government is willing to put into Unalaska.

"No idea," Tutiakoff says.

"I thought you said there was support coming?"

"They are *supportive*," he says of the government and Alaska state representatives. "That's the word they use."

I join the *Stratton* crew for drinks at Rat's, the local bar. The first shore leave is rowdy. I stay for three rounds. A young non-rate chugs a pitcher of pearly beer, stands, then vomits into the pitcher. The night is just beginning, and the howling wind raps at the windows. Rain forces its way through the window frame into my room at the Grand Aleutian Hotel, tracing water stains from past incursions. Everything seems to have fallen apart, to have melted, to have become untenable—or at the very least, extraordinarily difficult. Then again, this is Alaska. Somehow, this is America.

TOOLIK

NOT LONG AFTER KEVIN Williams annihilates a ptarmigan with the fender of our truck, we are skidding toward our deaths. Karma.

Minutes earlier, Williams had called over the VHF radio to announce the presence of our Dodge heavy-duty Ram supercab. "Four wheeler, northbound." Despite the tires being studded like shark teeth and having an enlarged and engorged gas tank, the pickup truck is the lightest and smallest on the haul road. "Four wheeler, northbound." These intermittent broadcasts are spoken into a landscape devoid of anything but a patchwork of gravel, dirt, asphalt, and ice, and groves of spruce clinging to the valley floor, with walls of mountain peaks dressed in a snowy plaid, closing in around us.

The ptarmigan has time to move; others in its flock do. But then again they are mostly dim-witted, lousy birds, and the fender is quicker than the straggler. The birds are everywhere in the north and yet in Fairbanks, which we departed from earlier this morning with a full tank of gas, bags of chips, and sandwiches, tourists from the "Lower 48" are hoodwinked into three-day ptarmigan hunts costing several thousand dollars; a similar experience can be had for the cost of a twenty-dollar rifle from the local gun store and a short drive anywhere outside the city.

An unintelligible voice broadcasts over channel 19. Then, ahead of us, an eighteen-wheeler approaches, behind it a cone of snow and dirt and smoggy exhaust. It careens past us and we enter its wake nearly blind. It is that easy to die in the arctic. In the blink of an eye. It may even be strikingly beautiful in the moments before.

We are some hundred miles north of the Arctic Circle and moving fast.

"Did you understand any of that?"

"I think he said something about a truck in his lane, I don't know . . ."

Only the sound of the wind and a tinkling of snow against the windshield, the tumbling treble of the studded tires, fills the cab.

"I think he said, 'Just to let you know.'"

Williams gets back on the radio and thanks the other driver. We drive on, wondering idly what he might've been warning us about.

The radios are the only form of communication, aside from GPS tracking devices moonlighting as text messengers, available along the haul road, known by its real name, the Dalton Highway, the only road in America that crosses the northern limit of wooded country. The Dalton Highway runs north from outside Fairbanks, some 250 miles west of the Canadian border, for 540 miles until its terminus in Deadhorse, on the shores of the Arctic Ocean, on Alaska's North Slope. A feat of engineering touted as the northernmost road in North America, it is considered a blight by preservationist groups and some Indigenous communities displaced by its construction, who still today—decades after it was built—are trying to recover from its disruption to nature and their cultural heritage. The truck motors swiftly through a pass.

There is a cottage industry here that supplies remote villages across the high north with access to the rest of Alaska: service stations, tourism, and necessities. This logistics chain, along the route bisecting the Alaskan wilderness, is meant not for the Indigenous or local communities but, rather, in support of the energy and defense sectors. The highway divides two wide swaths of national parks and has displaced countless residents, cordoning off traditional hunting and fishing grounds and rendering them inaccessible. It is the main artery for the extractive industries, with branches that veer off toward mining pits and exploratory drill sites. In some cases, it branches off toward property reserved for those displaced Alaska Natives.

TOOLIK

The Trans-Alaska Pipeline System threads along, under, and over the highway, delivering 463,232 barrels per day during its lowest month in 2024 to its port in Valdez. That made the Dalton Highway a project undertaken in the name of national defense. Built in 1974, the highway and the pipeline secured access to oil reserves and ensured the nation's role as a global player in the oil trade. The highway's namesake, James W. Dalton, also forged the northern network of early-warning defense systems called the DEW line.

In the way that America had seemed to revive, at least in writing, its efforts to ramp up military competency and assume a more robust diplomatic presence across the circumpolar north, researchers and scientists also began flooding back to the region in 2016. "Anything with 'the New Arctic' gets funding" from the NSF, one biologist who makes annual months-long research trips to the arctic tells me. Science diplomacy was until Russia's annexation of Crimea a mainstay in the arctic. With communications virtually severed between the largest stakeholder in the arctic—Russia—and the rest of the world since the invasion of Ukraine, including a decrease in the ease of access to arctic-going ships, scientists are exhausted and limited at a critical time for climate-change science. Studies on arctic climate suggest that restricting cooperation with Russian research stations, many of which have been banned from a major arctic monitoring network, hinders Western scientists' understanding of the global consequences of a diminishing arctic and further skews research interests toward North America and Europe. This hobbles predictive monitoring. Researchers say the climate at this point most probably cannot be fixed, no matter how many high-level discussions and forums are held on lowering global carbon emissions, owing to a lack of international cooperation. Twice the amount of carbon in today's atmosphere is also trapped in thawing permafrost, frozen in the earth's highest latitudes. Those gases are slowly releasing, and with Russia sitting atop more permafrost than any other arctic nation, researchers in the West are excluded from observing at ground zero the procession of climate change's impacts across the world. Scientists want to help, but it is too late. What funding there is for science—and there is never enough, even in peacetime—is ever more challenging to secure, as defense research takes precedence in the far north. There is no turning back time. The new Cold War will only get worse, in part because scientists failed to stop it and in part because the world failed to listen.

The neglect of the region and those working in it is visible along the Dalton Highway. It passes through a vast wasteland where all the infrastructure serves at the pleasure of the pipeline and the truckers driving to Deadhorse—Department of Transportation depots, oil spill cleanup sheds, shipping containers and trailers moonlighting as restaurants and roadside inns. Holes, known as thermokarsts, appear along the road and across the pavement, where shards of permafrost have suddenly melted, shifted, and given way. When the underground ice melts, the road goes with it. The thermometer can dip or rise 20 degrees Fahrenheit in the time it takes to drive a few short miles. Ravens and ptarmigan alight along the road as I drive with Kevin to the Toolik Field Station in the North Slope Borough. The name is an anglicized version of the Iñupiaq word for the yellow-billed loon, native to the region though, like much else here, it is endangered.

Toolik is one of the northernmost research facilities in Alaska (there are others along the Arctic Ocean shoreline, and at least one out on a berm in the Arctic Ocean itself), visited year-round by a revolving door of scientists undertaking projects to better understand the world's changing climate, despite their disconnect with half the arctic region. Many of the scientists working at Toolik have spent winters working in Antarctica, at McMurdo Station or the South Pole Station, often in the capacity of research technicians. Their job is to collect data for the scientists while the scientists are back home, in warmer climes, far from the arctic. There is a space-station–like camaraderie to living in isolation as they do. That was how Kevin and I found ourselves cheek by jowl on the nine-hour drive to the station; just the two of us, the resupply of food for the station including cases of Le Croix, and personal mail.

The road takes a dog-leg left uphill, then into a blind rise, as though we are driving into the sky. We drive over the rise and see the valley mousing below. We see the Brooks Range and the Arctic Ocean beyond. We see it is a false plateau, leading to another slight incline. We see what we could not understand over the radio. In the middle of the road is a jackknifed tractor trailer, its wheels skidding backward as the tires, wrapped in heavy chains, struggle to grip the ice. The entire truck is slipping backward. We make the turn at 80 miles per hour. There is no time to stop.

We are south of milepost 221, roughly sixty miles from the field station. The gap between the eighteen-wheeler's rear fender and the guard rail seems to meet in slow motion. Kevin cuts to the right and aims for the gap. The fender and tailgate close in. It seems impossible that we dart through, and when I look back as we pass the truck, the space we miraculously navigated has vanished, the road behind us severed by the jackknifed semi.

We arrive at the station after dark. Sustained winds of more than 50 miles per hour whip up snow bursts, and the trickling of white is illuminated by the headlights of a small snowplow operated by Joe Franich, the station's maintenance and support leader. Behind him is the quarter-mile-long driveway he has cleared for us; but it is already blanketed again by snow, his efforts undermined by the flurries.

"Stay to the left," Joe shouts to Kevin.

"Hit it hard and fast?"

Joe smiles and Kevin guns the engine. The rear of the Dodge Ram kicks a bit but steadies as we enter the station's lot, bordered by a dappling of buildings and warehouses and shipping containers illuminated by yellow and white lamps beyond which is nothing, a total whiteout. At least we are home and whole. There is warmth in the simple pleasure of arriving.

As I will soon learn, there are seemingly dozens of studies underway, from the way mercury is being cycled through marine ecosystems, to how archaeological and environmental archives will forever be lost as the planet warms, to what impacts can be expected from the loss of the Arctic Ocean's ice reflecting 70 percent of incoming solar radiation. Coastal erosion, changes in water density, algae growing on sea ice: These are just a snapshot of the many investigations underway in the arctic. Scientific studies of thermokarsts, for one example, are seemingly trivial but existentially important. Findings from those studies not only help civil engineers understand what the haul road needs to remain navigable year-round, but they also help climate scientists predict how much carbon is stored and being released by the permafrost.

Every study underway in the north is meant to support governments—indeed, humanity—in affecting changes in policy to prevent passing a "tipping point," beyond which the earth will continue to warm without stopping. It's something of a moonshot. Many of these projects, the foundations of the world's understanding of what climate change means, begin at stations like Toolik.

THE NORTH SLOPE, OR Arctic Slope, is by area the size of Nebraska; it is the only arctic tundra in America. February is the coldest month here, on average. Historically, the mean annual temperature is minus 22 degrees Fahrenheit, but it is warmer than that now—about 11 degrees Fahrenheit, breaking records each year.

It may be helpful to review how society arrived at this point. When nations pump carbon dioxide and other heat-trapping gases into the atmosphere—those by-products of an over-reliance on fossil fuels—global temperatures rise. That temperature increase is not only the antecedent to militarization; it is also behind the 100-year storms eroding coastal villages that were once protected by the annual sea ice. It is the reason that wildfires, once a rarity in the far north, happen with increasing frequency. What scientists in the arctic are trying to discern, for all their studies' nuanced complexities, is rather simple: What are we losing and what happens next?

It's true that scientists are often guilty, even by their own admission, of communicating uncertainty in their findings, making it easier for people to naysay the results. But this is largely a result of the nuanced nature of research findings, plus the massive Russia-shaped piece missing from the climate puzzle. Sometimes warning about a melting north is simply not enough. "It has also become increasingly acknowledged that science results should be effectively communicated to the public and stakeholders in terms of outreach and education," a research editorial published in *IOPscience* notes of arctic scientific literature, in the same obtuse language it sought to overcome.

To say the arctic is warming four to five times faster than the rest of the world—a grossly unexplained punch line—means little to those living elsewhere. It is an extreme that remains unfathomable and distant, removed from human consequences even as it causes real damage in the high north that ripples through the lives of people everywhere. It causes real damage in the global arctic as it does (or will) everywhere else. When the price of fish rises, when shipping and handling rates soar because ships are spending more on arctic insurance coverage, and when a movement of climate migrants creates greater domestic challenge, the reason will be those few degrees in temperature. Will Greenland's melting ice cap flood parts of Bangladesh? Will Cape Cod become prairie? Will sparrows make nests at the North Pole? Will all the disasters forewarned by climatologists be realized? We do not know. What has already come to pass is that the earth's climate first showed change in the arctic. The rest of the world then followed.

Seeking to answer all the unknowns are the scientists who return to or live in the arctic each year. "A flawless day," Ben Larson, a production cook at Toolik, says on the morning after I arrive. We are back in the four-wheeler, northbound, heading toward another stretch of barren landscape. Abby Jackson drives as Ben sits shotgun. We travel a short while to a turnoff north of the field station. Science is tedium, repetition. It is best not to do science alone in the arctic, so Ben is accompanying Abby, a research technician stationed at Toolik, as she retrieves samples collected by a three-story flux tower situated in the middle of nowhere.

Exiting the truck, Ben and Abby strap on skis. Ben is right; the day is flawless. The temperature is mild, and the sky is clear and blue, an expansive amphitheater blending into the snowy tundra before us and the sea beyond. Wispy breaths of wind grate granular layers of snow from the white-encrusted earth. The two seem to float as if on a nimbus cloud. The only variations in the scenery are man-made: Behind us is the pipeline; in front is the tower, part of the NSF's National Ecological Observatory Network, and a rectangular structure, like a child's playset on the horizon.

Our pace is slow and labored, and the trek is long. The snow here is different, wind-lashed abrasions that are harder, tougher than anywhere else I have been. There is a radiating connection to the snowy wilds one adopts after many months and years in the north. As foreign as the world seems from here, the reality of the impacts to the region are heartbreaking when, after a moment, it dawns on us that a few snowless winters may kill off many species. Eventually, one of those species may be human. As we ski, we come across a tuft of fur—perhaps a musk ox's—and take a moment to consider its origins. I recall a melancholic conversation I had once with Stefan Mikaelsson, the leader of the Sámi political party in northern Sweden: "Nothing will disappear and everything will spread. It's just a matter of time. When the time has come, it will be the public who takes the cost, the loss. It's not what can happen; it's what will happen," he said. He was specifically speaking about land-rights issues facing reindeer herders across the northern ancestral Sámi territory, which traditionally spanned from coastal Norway to Russia above the Arctic Circle. He then mournfully made those struggles universal. "I'm quite sure the people will exist but the traditional culture will be jeopardized. And arctic nations are treating them in the same way. They make it look like deserts, but eighty percent of the world's resilience comes from Indigenous populations, not only for the Sámi people, but for humankind."

GETTING TO AND WORKING in some of the most remote, enigmatic expanses on Earth to conduct their research, scientists contend with a slate of challenges both practical and political.

On the practical side, it takes time. There is a small contingent of scientists at the research station so late in winter, three of whom came from the University of Alabama on an NSF–funded project to study relationships in arctic spring-stream ecosystems. Getting here from the Lower 48 requires no fewer than three separate flights, before the long, meandering drive up the Dalton Highway. Then there is safety. After removing their skis, Ben and Abby must don helmets and harnesses to ascend the tower to collect samples, and they can never go out alone owing to the dangers posed by polar bears and inclement weather. A whiteboard back at the field station lists the departure time, destination, and expected return time of our trio, in case we go missing. Last, there are the collection methods. Abby has been trained to retrieve the glass jar filled with condensation without tainting the samples. She gingerly places the vessel into a box for safekeeping and an eventual return to Fairbanks with Kevin. When the samples arrive, the team in Fairbanks will use the data obtained from the samples to refine their models of forest structure and carbon cycling, eventually sharing the data and their findings with the world.

On the political side, before the conflict in Ukraine erupted into full-scale war, the arctic was a primary domain of international, collaborative research on the subject of greenhouse gases and climate change. As recently as 2021, Russia supported scientific missions throughout the Arctic Ocean, supplying teams with much-needed icebreakers, while also allowing international scientists the opportunity to study permafrost thaws in the country.

That is no longer the case. For the Arctic Council nations, this loss has been catastrophic. A variety of scientists, who do work similar to that being done at Toolik, have said the same thing for years. They all lament the death of their research collaborations across the arctic. "All my work with Russia in the arctic stopped," Dr. Tatiana Minayeva, the scientific director of Care for Ecosystems, based in Germany, told me. "I was working on the Russian Arctic only, hence my work on the arctic generally is over." If academic partnerships and collaborations will resume is unclear, forcing many scientists to shift to other areas of study virtually overnight. "It's all in abeyance," says Dr. Rob Huebert, a professor at the University of Calgary, whose work focuses on arctic security and sovereignty. "Any attempt to do just science on the practitioner level is over."

Collaboration in research among the arctic nations had until 2022 been ongoing since the Cold War, with the Soviet Union in 1973 signing the International Convention for the Prevention of Pollution from Ships, becoming a party in 1976 to a five-nation agreement on the conservation of polar bears, playing prominent roles in the International Maritime Organization and the World Meteorological Organization, and serving as a signatory to the 2018 Agreement to Prevent Unregulated High Seas Fisheries in the Central Arctic Ocean.

Even during the period of rising tension between Russia and the West caused by the conflicts in Georgia and Crimea, the Arctic Council managed to tread a precarious political line. But even in the most fraught periods, nonprofits were able to gather members of the Arctic Council, including the United States and Canada, to work out a proposal—for example, on the Central Arctic Fisheries Accord. But no longer. Getting the Arctic Council to meet around a table appears, for the foreseeable future, impossible. "The strength of the Intergovernmental Panel on Climate Change is that we bring scientists together from across the world," Dr. Jackie Dawson, a professor at the University of Ottawa whose research centers on the human and policy dimensions of arctic climate change, tells me. "With a nation missing, that's a huge perspective missing." The lack of knowledge sharing with Russia impacts everything: "The warming of the climate in the arctic doesn't stay in the arctic, it affects the entire world."

In March 2023, an Indian researcher published an article on the website of the Arctic Institute, a Washington, D.C.–based think tank, calling on India to use its G20 presidency to enable the resumption of research cooperation. "As an unintended consequence of the suspension of scientific cooperation in the Arctic, the Arctic research of the five Asian Observer countries—India, China, Japan, South Korea, and Singapore—also stands affected," the researcher wrote. For India in particular, arctic research is critical in helping scientists understand the melting of glaciers in the Himalayan region, which contains the world's largest freshwater reserves outside the North and South Poles. As for China, "China describes the Arctic as one of the world's 'new strategic frontiers,' ripe for rivalry and extraction," with a premium placed more on defense and less on economics and resources. Russia, while wary of Chinese ambitions, may welcome a strategic partnership in the arctic to balance NATO's expansion.

One project that was postponed, then canceled, was International Ocean Discovery Program (IODP) Expedition 377, an Arctic Ocean mission with plans to collect two 900-meter (half-mile) cores of sediment from the Arctic Ocean floor to help scientists better understand Earth's long-term climate history. The research mission was planned to run from August through October 2022, and it comprised a team of scientists from Norway, the United States, and Germany, among others, and was supported by two Russian icebreakers. One of the scientists on that mission, Dr. Katrine Husum, tells me its cancellation was a career low. Based with the Norwegian Polar Institute on Svalbard, Husum says she hopes that the international cooperation on which her research is built will one day return. "As researchers," Husum says, "it's really unfortunate, because the sort of knowledge gap that now exists is very vast." But it is not altogether unfamiliar.

When I reach out to Dr. Vladimir Romanovsky, a permafrost researcher at the University of Alaska Fairbanks, to ask him about the similarities between the atmosphere in today's scientific community and that during the Cold War, he tells me, "It was about the isolation of Russia at that time. In terms of scientific exchange, I would say similar. Maybe not as severe as right now. There were still some occasional contacts and there were several enthusiasts on both sides who would organize some contacts," Romanovsky says. "Russian scientists are very careful about communication with Western scientists and they definitely don't want to get any funding for fieldwork or anything else because they could get into the foreign agent status, which is very dangerous for them." I note that a colleague of his in the sciences weaponizes their data, withholding public release for up to six months, at which point it is outdated, in a small if effective protest against Russian and

Chinese researchers who support the Kremlin's global aggressions. "I am against this," Romanovsky says. "We are talking about using science as a weapon, and science has nothing to do with politics."

ABBY JACKSON AND BEN Larson return to the field station for lunch. There is a lounge off of the dining hall. Inside the room are a foldable tennis table, an ensemble of electric and acoustic guitars, La-Z-Boys, and a trash bin spray-painted with the letters B-U-R-N. There is a bookcase stacked with board games: Pictionary, chess, Trivial Pursuit, Catan, Taboo. But one game stands out: Pipeline. In this game, players are tasked with starting their own private oil company. Perhaps if fossil fuels are here to stay and target limits of greenhouse gas emissions aren't reached, a generation of children fearful for their planet will instead return to capitalizing on the destruction caused by their forebears. They will know nothing about international collaboration in the sciences. They could begin learning by playing Pipeline.

After lunch, Abby and I take some equipment and head onto Lake Toolik. The necessary tedium continues: It is time to measure snow and ice. I carry a clipboard, secured to my neck by a purple lanyard, as Abby jabs a long metal ice-core tool into the frozen lake. The tool whistles in the wind. Abby calls out every few meters—"twenty-seven centimeters, eighteen centimeters, thirty-two centimeters"—pulling out an ice core each time and placing it into plastic bags we have brought with us. The cores, owing to the unseasonable warmth, are frail and brittle. "I think we all have those thoughts about whether this makes a difference," Abby says. After a moment of reflection, after depositing more crumbly and slushy snow into a plastic bag, she adds, "I think Toolik makes a difference."

Just then the wind picks up. It is a warm breeze, and the way the snow scrambles over the ridges of the meter-thick frozen lake mesmerizes. For a moment it looks as though we are standing on a white-sand beach. The wind catches a loose plastic bag and it somersaults away. We give chase. Plastics pollution in the arctic is rampant even where there is little to no human activity, as here on the North Slope. Neither of us wants to contribute any more plastic to the ailing region, less the work of the scientist creates, rather than prevents, further damage to the environment; it's like the crew who came here from the University of Alabama and the moral quandary of the carbon-footprint cost of their sojourns to study a region impacted by that same toxic output. We secure the plastic bag and return to the station.

Back inside in time for dinner, we find there are kebabs with lemon slices and menial conversations about everyone's work that day. Abby makes lemon tarts topped with strawberries. In the scullery after dinner, I run into Alexander Huryn, the principal investigator on the University of Alabama project. He wears a flannel shirt beneath a dark brown safari jacket, with the breast pockets open to expose small pads and pens for making field notes. He looks at me through wire-rimmed gunmetal glasses perched at the hook of his nose, occasionally shaking his head at the realities of global health—how things seem unlikely to get better. "We have a lot of students who come up here saying, 'Yeah, we're studying climate change,'" he says as he empties his dinner tray into a large trash bin. "No, you're not; you're watching it. I tell all my students that if you want to collect data and add to our shared knowledge, that's great. If you want to have an impact, go to Congress."

The arctic is disappearing as the world encroaches. The best-laid efforts, by scientists, coast guards, and military organizations, are melting. Perhaps that is why so many are looking skyward. No longer are the arctic heavens solely home to the aurora, to the fantastical whims of solar flares and pitch-black skies. There, too, are the signs of invasion.

At night, I step lithely from my room at the research station, not wanting to wake anyone asleep in my trailer. An empty handle of single-malt scotch rests in the trash next to an empty handle of 100 proof whisky. A dose of arctic medicine.

I stand outside the tower of toilets—a raised platform with hollow shoots leading to a festering pit below—and look out above the station, momentarily stunned by the bright reports of the stars overhead. There is a gossamer screen of the aurora, and a great hole in the clouds opens to reveal a waxy moon. But the northern night—something almost religious in its imperma- nent wonder and grace—is disturbed. Though the sky is clear and the stars are bright, neither the crescent nor the starlight are the most captivating things overhead. The sky is alive, filled with satellites, trails of Starlink bea- cons which are bringing greater connectivity to those more regularly visiting this quiet place. The satellites are like curses whispering above, or outsiders, the celestial atolls of watchful invaders.

DRONES OVER DELTA JUNCTION

THE ARCTIC ANGELS, AS the U.S. Army's 11th Airborne Division is known, have trench foot. It is cold and windy and wet, and they are stuck outside in winter during the third Joint Pacific Multinational Readiness Center (JPMRC) training event in 2024. Inside two separate green field tents, medics are treating afflicted soldiers. Left untreated, trench foot could lead to sepsis and eventually death. There is a flurry of activity in one of the field medic stations. The entryway flaps briefly expose those inside to the cold outside. All the cots are full. Everyone seems eager to find an excuse to get out of the cold. It is warm inside, a respite.

From the very start, the exercise has not gone well. "Alaska weather got the best of us," a U.S. Army representative from the 11th Airborne Division tells me. Because of frigid temperatures accompanying wind gusts up to 50 miles per hour, the planned paratrooper jump and equipment drop is postponed. That night, from my room at a nearby lodge with cardboard-thin walls, I hear the yearnings and satisfactions of a noncommissioned officer and a local woman next door. I can also vaguely hear the low thrumming of helicopters on nighttime sorties once the weather clears, late into the evening.

In the morning, the simulated ground warfare resumes and there are more cases of trench foot. For nearly a month, more than 10,000 troops— mostly American but also some Canadian, Mongolian, and Finnish—have descended on the Donnelly Training Area, near Fort Greely, a slice of land along the Canadian border primarily used for cold-weather warfare training. It is the biggest military event in recent Alaskan history. It is also the largest such event since the 11th Airborne was reactivated as the nation's only dedicated arctic-fighting force.

The exercise overlaps with another, Arctic Edge, a biannual joint exercise hosted by U.S. Northern Command to provide further training across the Joint Pacific Alaska Range Complex to develop and test operational training in the trying conditions of the high north. The goal is shared by Canada's Operation NANOOK exercises, particularly the NUNALIVUT and NUNAKPUT exercises. Recurrent operational exercises promote relationship building, encourage interoperability between special-operations communities, and inform operational plans of existing gaps and associated risks. These types of activities contribute to the strategic messaging effort to deter encroachments into the North American Arctic, making good on the national strategy for arctic readiness and deterrence led by the Biden administration.

During the JPMRC training event, I exit the medical field tent and head to a combat medic station. The roads are congested with military vehicles, to the great frustration of local residents who find themselves regularly stuck behind long, snaking convoys of jeeps and flat-bed trucks, and the occasional Mine-Resistant Ambush Protected Vehicle. Many of them are painted in a palette more suited to the desert than the whitened clearings and snow-dusted Douglas firs. There are soldiers dressed in fatigues suited to jungle warfare, and they stick out against the craggy mountains of snow bulldozed into makeshift fortifications. One or two trucks are painted white; I am told this was at the behest of a squad leader, who bought the spray paint and did it himself.

My minder tells me that, in fact, the green camouflage is a benefit. White pants, green tops; better for blending into the tree lines where snow meets pine and bark. Perhaps, but what about open fields? The latest gear, he tells me as we pull off the main stretch of highway to a staging area, are the army's Cold Temperature and Arctic Protection System (CTAPS), which come in a single variety: woodland camo. These CTAPS are being tested by the 11th Airborne Division, which is based between Fort Wainwright and Anchorage. The five-layer system has been a success, says Sergeant First Class Michael Sword, the public affairs officer who escorts me around the battlefield. He tells me they are mostly effective. "There's a zone of what we call 'comfortably cold' and that's kind of where we try to live as best we can. So you're never super warm, you're never super comfortable, because as soon as you start moving, you'll warm up."

We pull into a parking lot outside the medical station. I am here because, like other military branches, the army has long suffered in its capacities in the arctic. The troubles the U.S. military has faced at sea in the arctic are not all that different from those they face on land: a lack of equipment, training, experience, and forward positioning. The army has also staffed the exercise with doctrine writers. Until this point, the army has been without definitive direction when it came to arctic combat, even for the division whose namesake was the region itself.

In the gravel parking lot, I spot Command Sergeant Major Chad Harness, the senior enlisted adviser for the 2nd Airborne Brigade Combat Team of the 11th Infantry Division, or the 2-11. He dismisses the delayed start as a nonissue, citing the fact that Russia and China lack the same advanced aircraft, and reasoning that if the U.S. Army is not making moves, neither is the enemy. He says the troops are learning a lot about their vehicles, including which seals and electrical relays are most likely to freeze and break, how to generally maintain their vehicles in the cold (which translates to learning to never turn off the engines), and to use communications equipment in these high latitudes. Nearby is a desert-brown Humvee with communications equipment. By a doorway, a portable satellite rests on the ground, pointing as close as it can to the horizon—a reminder of how far north we are—the trajectory being to the nearest satellites along the equator.

"With the technology that the army currently uses, we don't have direct line of sight, the further north you go, to a satellite to be able to take a shot for our satellite communications," Major Harness explains. "The farther away from the equator you go, the harder it is, so me trying to hit that if my shot is in this direction"—he wheels round to point with his hands like daggers—"I have to hit those mountains now. The chances of me hitting the satellite are slim to none."

I recall the pearl string of Starlink satellites I saw in the skies over Toolik, how the coast guard was testing it aboard the *Healy*, and how the Norwegian Coast Guard yearned for it. "It is only for our upper-technical operations," Major Harness says of Starlink. "I am able to get on the computer and stuff with it, but I'm not able to key a hand mic and talk to somebody across the battlefield with it." The hand radios present their own problems.

The L3Harris field radios are breaking down; they do not do well in extreme heat or in extreme cold—something the Swedish forces already know. Mechanical tubing in the vehicles is becoming too brittle; troops are using their personal lip balms to insulate and keep the lines from breaking. The weather is warmer than it was the week before, so morale is improving, even if the jury-rigging of equipment is taking a minor toll.

Not long before we spoke, Major Harness had returned from an Association of the United States Army (AUSA) event, and he realized that the arctic was coming into greater focus for the commanders, their subordinates, and the army more generally. The AUSA had found in 2023 that almost 80 percent of the army's budget was consumed by largely fixed operations and sustainment costs, leaving significantly less flexibility to fund modernization activities. Some of those activities included the acquisition of winterization equipment and arctic-colored battle-dress uniforms. "They are seeing [the value of arctic preparedness]; it's just a matter of the funding being allocated to it, or DEVCOM [U.S. Army Combat Capabilities Development Command] actually getting a hold of it," Major Harness tells me.

Inside the medical station, I am met by a group of medics. They show me the patient warming blanket, known as a "bear hugger." I point to the nearby cabinets that hold critical prescriptions and medicines. "They do make heated med chests. We just unfortunately haven't had progress getting those," one of the medical officers, Captain Impavidi, says. She takes me over to where a few orange buckets sit along the floor nearest a row of chairs. They are the same kind of buckets one might use during home renovation projects. Clipped to the edges are sous vide machines. "You might see them in your kitchen. The best practice right now for rewarming limbs, toes, fingers is to put them in thirty minutes of warm, circulating water from 102 to 108ish degrees Fahrenheit," she says. "We actually purchased these with government money. And put them in a Home Depot bucket, also purchased."

"The alternative is you're trying to keep warm using an electric kettle or a flame," another medical officer, Captain Colesar, adds, noting the field equipment already available to soldiers. "But you turn this thing on and it just works." No more trench foot.

THE AIRBORNE SOLDIERS OPERATING in the north have to contend with more than trench foot and cold weather. In 2024, the army began moving away from counter–Improvised Explosive Devices (IED) operations and, with the prevalence of electronic warfare seen in Ukraine, asked Congress for more flexible funding toward electronic warfare. Domain awareness across arctic battlefields includes cyberspace and the sky, and all the electronics therein, no matter how poorly a military's technology performs.

Critical high-tech gear simply will not always work in subzero temperatures where, among other things, batteries drain more quickly. Even howitzers freeze. America's long-standing inability to maintain or grow arctic capability goes beyond issues with equipment, though. Failings in domain awareness (underestimating the enemy's capabilities based on one's own inabilities, as the command sergeant major had earlier), communications (disparities in connecting battlefield units over satellite and radio in the Global North), and logistics (the freezing and breaking of mechanical equipment) are pervasive during the JPMRC exercise.

The U.S. military is forced to reckon with these challenges with concurrent real-world threats. Not days before I arrive in Delta Junction, the town nearest to the training grounds off the Richardson Highway, NORAD detected four Russian Tu-series strategic bombers operating around the Alaska Air Defense Identification Zone. Sorties by Russian long-range bombers through international and neutral arctic airspace are not uncommon. NORAD noted that it monitored the Russian movements through a "layered defense network of satellites, ground-based and airborne radars and fighter aircraft" and were "ready to employ a number of response options in defense of North America." But beyond monitoring Russian aircraft, there is little NORAD can do to defend against a concerted attack or a coordinated ground invasion.

NORAD is undergoing a multibillion-dollar modernization project to upgrade old, ailing, and inefficient computer and satellite networks overseeing the far-north defenses. The equipment is aged to the point where replacement parts—specifically transistors—are not in production anymore. As the successor to the DEW Line—that array of satellite dishes and radars and missile defense batteries set up during the Cold War to protect North America from over-the-horizon attacks—the North Warning System (NWS) relies on computers that lack the processing power or connectivity for video conferencing, let alone the ability to track emerging technologies and threats. The project came less than a year after the bizarre intrusion of a couple of Chinese spy balloons that wandered into and over the American and Canadian arctic regions.

In late January 2023, a suspected high-altitude surveillance balloon launched from China and drifted into Alaskan airspace. It hovered silently over the Bering Sea before the the military took action. NORAD scrambled fighter jets as the balloon crossed into Canada's Yukon Territory, traversing large swaths of the North American Arctic. The NORAD fighter jets tracked the balloon for days until it reached the South Carolina coast, where F-22s blew it out of the sky on February 4. It was not an isolated incident for the arctic: Six days later, another "high-altitude object" (the size of a small car) was shot down off the coast of northeastern Alaska. The incidents—described by U.S. officials as unprecedented intrusions—rattled nerves across the North American Arctic.

DRONES OVER DELTA JUNCTION

In Alaska, where locals are accustomed to military exercises like those outside Delta Junction or on St. Lawrence Island, but rarely foreign encroachments, the sight of jets intercepting unknown craft over otherwise untroubled skies underscored a fragile reality: The arctic's vast expanse was no longer a reliable buffer. From remote Alaska Native hamlets to NORAD's operations centers, questions mounted about why these incursions were not intercepted sooner—and whether they revealed dangerous cracks in North America's early-warning systems, especially in a state home to more weather-balloon launch sites than any other state nationwide.

One former senior commander in the Canadian Armed Forces and NORAD who oversaw the NWS told me that American intelligence had been tracking the arctic-bound balloon since it launched from China—that domain awareness was adequate, and the media coverage had allowed the United States to mask its capabilities. But real gaps do still exist. "Especially in the arctic, the two most useful things are chicken wire and duct tape," said retired Colonel Pierre Leblanc. "That's a problem. And that's why the system has to be modernized."

Whenever the modernization takes place, it will still rely on the foundational ground stations set up in the 1950s by James W. Dalton, in conjunction with more space-based monitoring assets and capabilities. When pressed further, the Standing Senate Committee on National Security, Defence, and Veterans Affairs in Canada noted that 38.6 billion Canadian dollars through 2043 would go toward strengthening its infrastructure, surveillance, and presence in the north. In the arctic, during a race against the climate and military incursions, two decades might as well be an eternity.

Then, Canada's Conservative Party leader and leader of the opposition, Pierre Poilievre, ahead of the election for prime minister, announced in February 2025 his plans for Canada's first arctic base since the Cold War, as well as plans for two new icebreakers and an overall shift away from North American cooperation in shared airspace. The plan was heavily criticized for diverting funds away from foreign aid; for placing the proposed base in Iqaluit, which is not optimal for either air or maritime operations; and for circumventing consultations with Indigenous organizations. He had also stepped in it the previous December, when he made a remark about a Canadian Arctic ambassador being a waste of money, as he believed the region was deserted.

America has bigger issues than its allies who traditionally have spent little on defense. As late as January 2020, the U.S. Department of Defense's Inspector General had warned of a severe gap in the nation's military ability to deter and defend against small, unmanned aircraft systems (UASs), a report which predated the balloon incidents said. That evaluation, too, fell desperately short. "Small," in Department of Defense vernacular, meant aircraft that are "commercially available and weigh under 55 pounds." Three years earlier, a group of military leaders had been assigned to "develop and submit recommendations" to help combat the threat of those aerial assaults. Ostensibly, the intention of this assignment was to make tactical improvements for soldiers fighting distant conflicts. There was little mention in the report, the "Findings" section of which is fully redacted, of a threat to domestic security. After those three years, though, the group and the Office of the Inspector General simply advised Mark Esper, then the secretary of defense, to appoint a military liaison to oversee teams countering small UASs.

But a UAS incident is not strictly relegated to foreign combat. A UAS incident covers a wide spectrum of possible assaults by a foreign military or domestic enemy. If a drone flew over or near a military base, facility, convoy, ship, airfield, or other defense infrastructure, that would be a scenario the report was glancingly meant to target. Whatever the working group recommended did not seem to apply in the case of the balloons. The report, it seems, was too narrow in scope, and its findings were seemingly never implemented. At the very least, it proved to be of little value to arctic operations and to Alaska specifically, where more military assets are placed than one might first see.

When a map of U.S. military infrastructure and installations in Alaska is displayed, one typically sees only three dots: Eielson Air Force Base, Kodiak Air Station, and Joint Base Elmendorf-Richardson. What one does not see are the six other major military installations, the roughly fifteen NWS radar stations, the fourteen Long-Range Radar Sites, and a handful of smaller forward-operating locations and research sites, like a permafrost research tunnel. Even the navy's Southeast Alaska Acoustic Measurement Facility (SEAFAC) goes unheralded. Alaska is a petrostate. It is also a military-industrial state. It's one big defensive position, falling into neglect.

ON A RIDGE OVERLOOKING the battlefield, smoke rises from various out-posts. An artillery battery of M198 155-millimeter howitzers is set up near a clearing along a river, and if I strain, I can see the movement of some troops dressed in white pants and rain covers, gadding about near brown military vehicles below. Two men from the U.S. Army Aberdeen Proving Ground in Maryland have parked their red pickup truck on a switchback that leads into the valley below. They load a tennis ball onto a quadrocopter drone. The drone takes flight with its payload, zipping toward the troops. "The thing is, like, it's not really a game," Sergeant Michael Sword says. "If you treat this like a game, that means you're going to treat the real thing like a game. The stakes for this thing are real."

The two drone operators agree through weak smiles, their gazes affixed to screens and their next kills. I watch on a small tablet as the drone descends on a group of the 11th Division, and it releases its tennis-ball payload, simulating a drone strike that kills a handful of the soldiers. The operators laugh, watching the soldiers scurrying below, trying to shoot the drone down with a shotgun and dummy ammunition. "The people that do well here usually do well anywhere. They have more of an opportu-nity mindset instead of a fixed mindset," one of the drone operators tells me. "When the towers fell, our focus was on Iraq and Afghanistan. Now it's, like, the focus is up here, because there is a much-needed and much-overdue focus."

The drone returns and is loaded once more, before being sent out again to target the same troops. "This place really should be an all-volunteer unit. But it's gotta be built up over time. Because culture takes years of change," says the operator, who gives his name only as Bruce. "Unfortunately, not many people are willing to do that because most people are only here for three years. There are only a few of us who stick around for longer. By the time someone figures out their job, they are gone."

It is not clear whether it is to Washington's credit or to its detriment that it has finally regained its frenetic interest in Alaska, in the arctic. It is too early to tell. At the very least, combat troops are gaining the necessary experience in things as simple as remembering to zip their gloves inside their pockets when not being worn, lest they lose them and a hand or two. They are also learning that snow does not protect against bullets and munitions the way sandbags or sand dunes might. That's how, down in the valley below the drone operators, I found a soldier from Texas using his shovel and wooden detritus to fortify his position, packing snow into something like a toddler's version of an igloo.

I keep my eyes skyward, aware of the drones overhead. First Sergeant Justin Warner, of the 1-11, spots one and tries to shotgun it down. Then there's a whistle, a howl, and several explosions. Four large smoke grenades explode a few yards away from us. Only two of the soldiers—the company First Sergeant Warner and his aide—survive the artillery strike on their position. Everyone else is dead.

Looking around at his troops as they gather in the formation of simulated death and march off the exercise, Sergeant Warner hangs his head. The drones could have helped call in the artillery that killed his troops. There were things they still needed to learn, but above all, they needed a cohesive strategy, a guide. "This is a cultural thing," Sergeant Warner says.

SNOW AND ICE IN popular Western culture are always portrayed as villainous. In classic and modern literature, in cartoons and storybooks, it is the cold weather that signals terror and wanton dereliction. Rarely is cold weather embraced or rejoiced. The cold is usually something to overcome, a burden, never something through which one might strengthen. A newcomer to the arctic will always be bombarded with questions about attire, and asked whether they are prepared for the winter, since the cold is the number one killer. The Arctic Angels would do well to remember, however, that vilifying and seeking to avoid ice and snow are not their only choice; they can also use it to their advantage. An army handbook published in 2020 describes the snow and ice as an advantage, through which soldiers can use frozen bodies of water as avenues of approach; create makeshift fortifications from compacted snow; conceal movements in deep snow; use specialized vehicles designed for snow travel, rather than those traditionally deployed to the Middle East like Strykers; and leverage the harsh conditions to hinder enemy operations in areas where they are not properly equipped for cold-weather combat.

Some have gone as far as suggesting the destruction of the polar ice cap altogether. Weaponizing the arctic by, with, or through science has long been a curiosity of researchers. The arctic has also been viewed as a region not only for weapons testing, like nuclear bombs in the Russian Arctic, but also for those bombs that could themselves be used to alter the climate. Dr. Harry Wexler, then the director of meteorological research at the weather bureau in Washington, D.C., wrote in *Science* magazine in 1958 that a series of proposals to modify weather were "impractical or likely to produce cures that are worse than the ailment." Take, for instance, the proposal for a handful of well-placed hydrogen bombs beneath the Arctic Ocean sea ice that could halve the amount of heat dispersed into the atmosphere. News of Dr. Wexler's article, "Modifying Weather on a Large Scale," was buried twelve pages deep in the *New York Times* on November 2 of that same year, and appeared next to an article describing Christchurch, New Zealand, as a key city for work in the South Pole. All angles of approach to the poles were hot topics back then.

Dr. Wexler's rebuttal to the thinking of the day was simple, though not without its caveats. Ten ten-megaton bombs would "throw enough steam into the polar sky to blanket the Arctic Ocean with an ice fog," the *New York Times* reported. "Since winds and ocean currents from warmer regions keep carrying heat to the Arctic Ocean, interruption of its escape would inhibit the formation of new ice." Back then, the northerly polar ice cap lost three feet each summer. Dr. Wexler said that the ice cloud would "accelerate greatly" the ice cap's shrinkage, and would create an ice age for those living in the mid-latitudes.

Today, the notion of harnessing the power of the north for good or bad has taken a different turn. In 1968, the army's Engineer Research and Development Center (ERDC) excavated a tunnel through layers of permafrost outside of Fairbanks. After the tunnel was largely abandoned, renewed interest since 2014 has left the project in perpetual need of more funding to keep up with the demand by scientists, many of them from the Pentagon or its contractors. That permafrost tunnel is run by the ERDC's Cold Regions Research and Engineering Laboratory and is technically on the base of Fort Wainwright. Militarization and defense research projects abound across the state and its territorial waters. But at the tunnel, there is more focused research on how a warmer future could be a boon to national defense and used as a hostile deterrence.

On a visit by Defense Department officials in 2023, a military team trained for a mission during which they were tasked with finding biological material they might weaponize. Would smallpox or the Spanish flu resurface? Could it be the source of another pandemic, natural or man-made? Therein lies a superfluity: the nation's leaders want to deter others from committing harm to the nation, but are willing to harness that same violence in order to wield it. What is deterrence if not passive aggression? The militarization of the Russian Arctic comes in response to perceived aggression by NATO states, which comes in response to other perceived threats and so on and so forth, ad infinitum.

A senior defense intelligence analyst explains, "The arctic is the only place on Earth that, in and of itself as an environment, can be weaponized. The only environment that can be weaponized. How? By virtue of its physical location, its position with respect to the sun, its magnetic properties and its ability to project a disproportionate effect outside the arctic [from] within the arctic," the analyst and scientist working for the Department of Defense tells me. They were hired by the U.S. Northern Command, which oversees control of Defense Department homeland operations, to figure out what new pathogens could come out of permafrost and how those could impact operations or, as the analyst says, "how they could be weaponized."

ANGELS IN THE DARK

DISQUIETING DARKNESS SO FINAL it is like the world inside a coffin. Yet inside the funeral home in Anchorage, nearest a post office and within walking distance of America's Diner and a big-box store along a snow-blanketed sidewalk, it is well-lit and bright. Delilah's been a funeral director since she was in her early twenties, when we first met a decade ago. Her mother's being from Nome and her father's being from the Lower 48 split Delilah's allegiances between the Iñupiaq and the white man, though she firmly associates with the former, her people and her tribe. There is little extraneous space in her world—a mixture of the maudlin and the formalin—which she navigates in formal attire in shades of dark grays, deep blues, sometimes threadbare browns. Delilah wants to be cremated (she trusts no one else to do her work) and despises organ donation. Why waste a perfectly good chance to embalm, to preserve what remains?

Delilah readies one day to bury a woman whose eyelids are smudged, first the makeup then the epidermis itself, crinkled by a concerned loved one's final altruistic overreach. Delilah works in the back of the funeral home. Autopsies are not performed here, and in fact they make her work nearly impossible, since the circulatory system is dismantled in the process. She cannot reclaim what has been taken, so she does her best when those cases arrive, the organs rummaged through and placed neatly in a bag inside the body cavity, like the arctic has been: examined, violated, and shoved aside.

A U.S. Postal Service mail carrier walks into the mortuary and goes to the receptionist with a bundle of letters. He speaks momentarily with the receptionist, out of earshot of Delilah and the mourners elsewhere. The receptionist has no outgoing letters or packages. "Oh, come on. People are dying to give me mail," the mailman says before departing.

Outside, among the eateries and megastores, and farther still into the Chugach Mountains, is a similar, though larger and more kinetic, world: the Alaskan wild. The midnight sun with its diaphanous light sets somewhere between dusk and dawn. Towering monoliths of ice and floe wedges, one atop the next, are a staircase toward a distant forever. Mysterious creatures, some with unexplained tusks for unknown uses, are terrifying in their might yet life-affirming in the sustenance they provide. Marrow-freezing cold is an impediment to those very affirmations; electrons collide with nitrogen or oxygen to produce, respectively, the northern lights of neon umber and emerald, and sometimes pastoral blue. When these colors mingle, they create violets, fuchsias, a doleful white that appears suddenly, lingers, vanishes.

Delilah knows well this arctic and its so-called Angels. Many pass through her doors on the way out of that wilderness. The division, formally known as the 11th Airborne, is setting itself up to become the "tip of the spear" in the arctic—America's go-to permanent arctic force. Other military units train in the far north, but the Angels live here though their home is not always a place of tranquility, beauty. More than the wounds of combat and training bring them to Delilah. Psychological warfare and the impacts of the high north on those who serve here are real and pervasive, and have seemed at times insurmountable. Delilah regularly bears witness to the results.

Delilah is persistently busy, peering into the faces of mostly young males who come through her mortuary. Sometimes they are gunshots; other times they are fentanyl overdoses, treated as such but suspected as suicides. Their deaths highlight the ongoing need for comprehensive mental health support and the importance of sustained efforts to address the unique stressors faced by soldiers stationed in remote and harsh environments like Alaska.

THE ARCTIC ANGELS ROTATE through this mortuary with worrying frequency. In 2021 alone, nearly twenty soldiers from the 11th Airborne took their own lives. Commanders said they saw none of the typical warning signs of suicidal ideation: depression, the offloading of personal effects, social isolation. Alarmed by the rising suicide rate among active-duty personnel—many of whom had never seen combat—Alaska Senators Murkowski and Dan Sullivan pushed for change in 2022, successfully lobbying Congress to increase funding for military suicide prevention. Their efforts also led to a Department of Defense investigation into suicides at remote installations like those in Alaska, where the harsh conditions amplify soldiers' struggles. Military leaders then began working toward a more holistic and welcoming environment that fostered fellowship and camaraderie against the long Alaskan summers with its midnight sun, and the long Alaskan winters with its temperatures below minus 10 degrees Fahrenheit. They renamed the division the Arctic Angels, apparently failing to predict the morbid inferences soldiers would come to make (that their posting would damn them to an early grave), not only as part of a rebranding of sorts but also as a way to tell service members that this was where they belonged. That they were special. That there was meaning to serving in the harsh and forgotten North American Arctic, a national appendage far from the Forever Wars.

Originally activated in February 1943, the 11th Airborne fought across the Pacific theater during the Second World War. They were distinguished, recognized as the Angels for their speed and lethality, their ability to become action-ready at a moment's notice, and for their means to quickly organize into a fighting force. Among the many honors the division received, two privates—young soldiers—were awarded the Medal of Honor posthumously. The division was then stationed in Japan to oversee the occupation forces before returning stateside, and in 1965, it was rather unceremoniously deactivated, the troops disbanded to elsewhere. Their history seemed to die there.

The change in nomenclature for the 11th Airborne to the Arctic Angels and Arctic Wolves carried momentous implications. The forces stationed in the north were there to stay and would become de facto experts on cold-weather warfare year-round. Rebranding efforts like this had not historically proved entirely effective. The problem, it seemed, this time at least, was in the army's approach to plummeting morale: It refused to do anything unfamiliar. It had tried the same "reflagging" strategy on the same division in the months leading up to the war in Iraq, but that changed nothing about the unit's morale.

But more than half a century later, on June 6, 2022, on the seventy-eighth anniversary of D-Day and the Allied invasion of France, the division was reactivated both at Fort Wainwright and at Joint Base Elmendorf-Richardson and renamed in response to the suicides a year before. "Between the end of the Vietnam War and 9/11, we were basically a hunting-and-fishing brigade in the far north; it was real demoralizing," one soldier was quoted as saying in 2004 after the 172nd Separate Infantry Brigade was similarly reflagged. That unit was about to become the army's third Stryker Brigade Combat Team, and eventually become part of the Arctic Angels. They were then identifying with their Stryker vehicles as their specialty, despite being based in Alaska, where the vehicles were useless anywhere other than for overseas deployments. The Stryker became their calling card. Then, at the 2022 ceremony, they were christened the Arctic Angels, harking back to their erstwhile import. Their mission was to be the army's only permanent arctic defense force, to defend critical infrastructure across the north and be capable of outmaneuvering, outshooting, and outlasting adversaries in the frigid climates of the northern frontiers of our planet.

The suicides persisted despite that rebranding. After the 11th Airborne Division's reflagging to their new identities as arctic and airborne soldiers, the unit ditched its Stryker attack vehicles from its time as the air-assault division, dismantling the unit. The vehicles worked well in Iraq; they did not perform well on snow or ice, and they led to hours of frustrating maintenance for the soldiers. Removing the vehicles was also meant to boost morale.

"We are good at what we do, however we are just not filled with joy when we go do it if that makes sense," one specialist tells me through an encrypted messaging app during a training exercise outside Fairbanks. "They were trying to boost morale by rebranding us so to speak. However, shocker . . . no one cares about that stuff. Everything has stayed the same morale wise. I was unaware that getting rid of Strykers was supposed to boost morale as well. To my knowledge, it is purely based on boosting Arctic combat effectiveness." The rebranding and the removal of the Strykers were meant to do both.

THE ISOLATION, EXTREME WEATHER, and grueling pace of operations create a unique crucible of stress, leaving service members vulnerable to despair and suicidal tendencies. The urgency was clear. In 2023 alone, 523 service members in the army died by suicide, marking a 9 percent increase from the previous year and reflecting a disturbing trend that has climbed steadily over the preceding two decades. Young men in the military are more likely to die by suicide than their civilian peers. But it is not likely that combat-related PTSD is the primary driver. Instead, it is often those who have never deployed who are most at risk. The 11th Airborne has faced its own significant challenges with soldier suicides, with the division reporting seventeen suicides among its ranks in 2021, which prompted unit-wide introspection and a comprehensive response to address the crisis.

"To be fair it's not an easy place to be. It's colder than any other base in America. The jobs we do are tough and require a lot of resilience," one junior enlisted soldier in the 11th Airborne's 1-11 tells me. "There's no nice way to put it other than some people can't handle it. Not to say I haven't had times up here where I'm depressed and in dark places. I wouldn't completely blame the army for the suicides."

Along with its rebranding, in early 2022, the division launched Mission 100, a program mandating wellness counseling for all soldiers to provide early intervention and destigmatize mental health care. But the situation remains complex. Between October and November 2022, four soldiers died in suspected suicides, underscoring the persistent challenges the division faces. But by July 2023, this initiative and cultural shifts had shown promising results, with a notable reduction in suicide rates. As of mid-2023, there had been one suspected suicide, a significant decrease from previous years.

The state and federal initiatives championed by Senators Murkowski and Sullivan laid the groundwork for the Center for Alaska Native Health Research (CANHR) to receive its first grant to work directly with the Arctic Angels. Just a short drive from CANHR's offices at the University of Alaska, the division's 1st Brigade became a proving ground for CANHR's approach, rooted in partnership and adaptation. Drawing from the Qungasvik model—designed to bolster mental health and reduce suicide in Alaska Native communities by connecting youth to the culture and community they find themselves part of, no matter how remote and unforgiving—CANHR enlisted military officers to collaborate on strategies tailored to the army's unique challenges. Together, they sought to address the symptoms and the underlying isolation, monotony, and stress of life in uniform in one of the most unforgiving environments on Earth.

DR. JAMES MORTON JR., who worked alongside CANHR and the U.S. Army, did not fear a challenge or the cold; he embraced both. He was the kind of Special Forces Operational Detachment Alpha soldier in the army to remove his gloves while trekking through arctic Norway—not out of machismo, but because his hands were too clammy. Skiing, mountaineering, extreme temperatures, and high altitudes—his deployments around the world had all the trappings of an action-film blockbuster. But life in the Special Forces, specifically the Airborne Division, often relegated to the arctic and subarctic on training missions and exercises, was less than glamorous, despite the depictions of serene landscapes pictured on the postcards sent home.

After earning degrees in systems engineering and international relations, Morton left the military for academia. He earned a master's degree as a mental health professional counselor and his doctorate in counselor education and educational psychology. He went to work at the University of Alaska Fairbanks' Center for Arctic Security and Resilience, where he led research projects about the adaptability of Special Operations missions in the arctic. That work could prove to fundamentally alter the way the American military operates in the high latitudes, such as the Arctic Angels or the Navy SEALs waylaid on St. Lawrence Island.

Morton is the principal investigator at the Military Health and Readiness Consortium, where he envisions an evidence-based, cognitive science–oriented solution to the Angels' identity crisis. He co-constructed in partnership with senior leaders of the 11th Airborne a structured lesson plan to reeducate commanding officers and noncommissioned officers on how to instill purpose, meaning, and better life practices in their subordinates. Morton and his colleagues looked to Alaska Native cultures, adopted what had worked for some communities along the edges of the Bering Sea that had faced high rates of suicide. He believes it could work. The first classes, which began in the spring of 2024 at two bases in Alaska, and the responses from some noncommissioned officers who attended the courses, proved promising, hopeful even. But even as the program moves into its next phase, as it branches out to reach more regiments within the 11th Airborne, the changes remain to be seen though statistical analysis has shown progress.

The Purpose Driven Leadership (PDL) course, which uses basic cognitive science in reverse to effect change before the change would otherwise be necessary, is aimed at noncommissioned officers—part of the mid-level tiers of command leadership who on a daily basis interact with and train combat soldiers. Among these leadership ranks are the team leaders, the squad leaders, and platoon sergeants. Much like educating Alaska Native youth on their roles as young leaders in a community, the program develops the relationship between a soldier's perceptions of leadership style and the purpose of military service that may mitigate the risk of soldier suicide.

Morton's three-hour course begins with a focus on building a specific identity for the Arctic Angels. Their test groups were two battalions that had recently completed winter exercises across Alaska. When Morton and his colleague, Benjamin Trachik, a researcher at the RAND Corporation, spoke after I first met Morton in Fairbanks, the 11th Airborne had just finished a joint deployment and exercise along the Aleutian Islands, part of an increase in efforts to project American military reach across Alaska and Hawaii as a form of deterrence through "force projection." This was only a few months after I embedded with them for the JPMRC training program.

"They can think, this makes sense. This is how we do the fight!" Morton says. "So, enlisted soldiers, E1–E4, they are really kind of struggling with, 'How do I be in the army? Who's my de facto mom and dad, and how do I fit into the fight?' But when they get to the larger collective exercises or activities, they can see, 'Okay, this is how I fit into the larger picture.' It's a bit harder because we are still in an adolescent mindset, we are still learning how to find our place and who are we." Developing a sense of responsibility and ownership is where the course helps noncommissioned officers (NCOs) better answer those questions and instill those feelings of unity, providing a sense of duty and place in the army's only arctic brigade.

"What's different about what we're doing is [that] we're saying what's unique about this population that is naturally protective," Ben Trachik says. "Naturally protective, meaning, what does the literature show us if you have the presence of this positive trait? You are less likely to go on and make a suicide attempt."

The program is still in its infancy. When I follow up after they have completed the first trial of the study, and I note that there were, during the course of training the two battalions, a few more suspected suicides, both Morton and Trachik solemnly nod.

For now, the PDL course is a series of slides in a PowerPoint presentation paired with open discussions. One of the first slides quotes Brigadier General Bill Mitchell's testimony to Congress in 1935, during which he said, "Alaska is the most strategic place on earth." Adjacent to the quote is a map of Alaska that's overlain with a heart and the words "Military Heartland." After a brief history lesson about the 11th Airborne, whose significance in the Pacific theater during the Second World War is often lost on a generation that no longer reads Stephen Ambrose, the discussion moves to encouraging the NCOs to describe their experience doing a polar plunge—diving into freezing water as part-initiation, part-practical training. Recalling the experience is aimed at reinforcing that Arctic Angel identity, the fierce wolves who patrol the far north. At best, the course is about reiterating notions that come secondary to allied nations that are already operating frequently in their arctic regions: They are alone, without the likelihood of immediate backup; that things generally take longer to do than they do in warmer climes; and that everything breaks at minus 40 degrees Fahrenheit or below. Rather than being seen as a punishment, as being in a forsaken place to serve, a region forgotten by the rest of the nation and Washington, serving in Alaska should be seen as a mission- and life-affirming assignment.

With broader strokes, the same mentality could be applied to any discussion in Washington about the arctic and about America's role therein. It is not simply a place of woe, of climate disasters and the people displaced by them. Nor is it only a military staging ground. It is, rather, a place of benefit to the world and the nation, a place that offers much in the way of life lessons and economic sustainability, if not outright national defense. If treated right, if nurtured, and if a culture of understanding and reason toward the region is fostered—one that illuminates in the minds of the nation's leaders in Washington and in corporate America—the far north, Alaska, and the arctic regions at home and abroad can provide much more than simply income and natural resources. Then the country wins. The people win. Then thrive.

DELILAH, THE MORTICIAN, IS beautiful—pierced septum, bespectacled, and tattooed—and would perhaps wear something warmer than a blouse, a suit jacket, and a pencil skirt were it not for another recent, unseasonable heat snap. She confers with the family of the woman whose makeup needs a touch-up before leaving with the bereaved for the cemetery. Upstairs, where the hallway smells faintly of potpourri and hand sanitizer, its walls adorned with Alaska Native artwork—the slaying of a seal, the aurora over musk oxen and *kawartarwiks*—there is a room lined with choices. Urns of wood and paper and industrial plastic. Flag boxes and challenge coins for the Armed Services members who pass through. An entire section of wall is devoted to those who served their country, in a state hosting thousands of veterans and active-duty service members. Every day more service members arrive in Alaska—a state that could consume Texas twice, with room left over for a helping of Rhode Island—some responding to threats at sea by China's ad hoc maritime militias, others to airborne incursions by Russia and the unknown. It is a solemn reminder that much of what we know of the arctic today was founded, and persists, on the shoulders of military industrialization, ambition, expansion. That it has gotten us only so far.

A raven pokes around in search of food in the parking lot outside the funeral home in Anchorage. The pavement is bracketed by plowed snowbanks muddied by recent rain. Flags of four of the five military branches fly on poles placed in intervals along the roadway—a banner for the U.S. Coast Guard notably absent. Delilah received a caller once who complained that the flags were too wind-sheared, tattered, in need of repair and support. The same could be said of those military units, the people here. The region writ large. Delilah remembers shrugging in response. What could she do against forces stronger than her?

In Delilah's absence, the funeral home echoes with blitzing telephones and a receptionist whose demeanor is neither gentle nor kind, but matter-of-fact and upbeat. Life moves in all directions around death and upheaval. It's a river flowing in grand perpetuity, if sometimes momentarily frozen, before rushing on, across an arctic nevermore. Delilah and I stay in touch. For months after we met in Anchorage, long after Dr. Morton's trial program has ended, my phone occasionally buzzes with a message from her. It is often late in the evening in Anchorage. "We've had two active-duty deaths this week," she texts me one day. "Both self-inflicted gunshot wounds."

THE LAST TREE

IN THE SHADE BENEATH a lone tree, the drone takes flight and we silently pass cigarettes between us. It is below 20 degrees Fahrenheit, and the drone ascends the ruffled western facade of Sukakpak Mountain's brilliant Skajit limestone bathed in orange light. Each breath of smoke freezes in the lungs. The side of the Dalton Highway is quiet, save for the frequent eighteen-wheelers that appear as headlights in miniature whirlwinds of snow dust. Two arctic field-support specialists watch and smoke as the third peers through goggles to see what the drone sees: a fanged ridge crisp with snow. What adjectives to ascribe it? Proud? Dignified? Austere? The *Mighty* Brooks Range? No. Those are words for us, not nature. Nature has its own language.

Thirty miles south of here, somewhere outside Coldfoot (120 miles above the Arctic Circle, so named for the gold miners who reached there years ago, then out of fear or frozen expectations, turned back), and the last place to fuel and eat before reaching Deadhorse and the Arctic Ocean, the boreal forest ends. Beyond the ridge, what the drone can see is distant mountains. And beyond them lie the boreal of the Canadian Arctic, the Yukon, the Northwest Territories, and Alaska below. As the drone reaches the mountain peak at 4,459 feet, the operator can see the boreal's gradation slowly diminish, getting smaller as it stretches toward us into the North Slope Borough. We were at the precipice, where foliage cedes to icy expanses. This human thing, where we strive for normalcy through obstinance, leaves its mark here; there's nothing in between to temper the relationship we have to the north—no compromise, only struggle.

The boreal forests south and farther east can be imagined as something of a protective curtain, the shade it provides also a vast carbon storehouse. Collectively, the forests store 30 to 40 percent of all global terrestrial carbon. Wildfires in Canada's northern forests, which seem to happen now regularly every summer, release carbon dioxide that warms the region and promotes more frequent fires. Since just after the turn of the twentieth century, some 70 percent of all tree-cover loss occurred in boreal forests, largely in Siberia and Canada. Fires in the Northwest Territories just shy of the Arctic Circle, ringing the Great Slave Lake, in 2023 engulfed a landmass roughly the size of West Virginia. Arctic forests are burning faster than they can regenerate. Full of conifers, they are the world's largest biome, stippled with pines and spruce and larch. Across these unsettled expanses, though slowly crawling northward as the planet warms, air force flyovers and mining roads are becoming more frequent points of greater and ever more important consternation.

Our lone tree offers its shade and the sun, which hangs like a lampshade on the horizon to the south, coaxes great spindly shadows to grow. The cigarettes are stubbed out, placed in a plastic bag for disposal later. We scramble back into the truck when the drone returns. On the radio, we overhear a conversation between truckers.

"Heading off to the Virgin Islands?"

"Yup."

"Hot damn! Well, you enjoy yourself, and we'll see you when you get back, Casey." There always seems to be a mass migration here, of those coming and going and returning. Conversations are clipped. The VHF radios reach only so far, sometimes terminating at a mountain pass. Reception is a lifeline. Absent reception, the closest thing to a lifeline, a connection to anything apart from the land, is our contacts with each other. We sit in absolute silence, consumed by ourselves, indifferent to a place we have frequented, made a second home, choosing now to look away out of shame or apathy, or a toxic mixture of both.

CONSERVATION IN ALASKA HAS two sides. On one, there is the natural world, of spruces and species. On the other, there is our imposition of the material world, our need to sustain what we have built and how we have learned to survive. The sides clash on lands inhabited by Indigenous peoples; it's the wants of the few versus the needs of the many. More frequent yet are worries expressed about the lack of Department of Transportation workers along the haul road; or fears of the pipeline one day bursting; the sheds and warehouses along the road that house cleanup material seem wildly small, as though they are meant as theater and that what is contained inside could do very little to avert ecological disaster. Or, it's feeling the cost of living is too high; or that the Permanent Fund Dividend checks granted to Alaskan residents (and born from that very pipeline) are insufficient. Priorities are skewed.

The debates in Washington are about whether the nation needs more oil, or whether to pave a road through the Gates of the Arctic National Park and Preserve, deep into untouched wilderness, so as to access world-class resources of copper, zinc, lead, plus the associated silver and gold. The road project is needed to increase job opportunities and encourage Alaska's economic growth, but there is little consensus between the state capital and the Alaska Native communities. "We deserve the same right to economic prosperity and essential services as the rest of this country and are being denied the opportunity to take care of our residents and community with this decision," North Slope Borough Mayor Josiah Patkotak said in 2024, when the Biden administration blocked the project. But Voice of the Arctic Iñupiat, a pro-development advocacy group, was satisfied—at least for now—"that caribou were heard over cash."

Their respite lasted one year, then President Trump overturned Biden's ruling. Conflagrations over natural resources, and what they give and take, will continue to divide as the north opens up, as the planet looks toward a green transition still requiring a dive beneath the surface of the planet. Near

Yellowknife, in Canada's Northwest Territories, rich deposits of lithium, used in electric-vehicle battery manufacturing, have drawn the interest of at least seven companies seeking to explore the terrain's potential. The U.S. military also needs antimony, found in chunks of stibnite, a little-known ingredient used in flame retardants, solar panels, and missiles. Right now, China is the U.S.'s biggest supplier of the hard rock. Australian companies are looking to Alaska to open such mines in the coming years, in an effort to lure customers away from China.

It is a compromise America needs to make: to recognize security and stability in all its forms, from within and without, to look beyond the horizon, as it does with radar and satellites, to better protect the natural world. We need to similarly protect the resources that are often taken for granted, to give the same weight to the natural world as we do to protect national sovereignty and, by that same token, its longevity.

OUR REST STOP AT Coldfoot offers two menus, one with what's on offer, one with what is explicitly not. Not far from the rest stop, there was once a tree. Through the impulses of whatever tourism or conservation bureau, a sign was placed beside this tree near the headwaters of the Dietrich River: "Farthest North Spruce Tree on the Trans-Alaska Pipeline: Do Not Cut." Superlatives are frequent, and perhaps frequently offending, in the circumpolar north: the last this, the farthest that. Is the value here only in its finality? The spruce was roughly 273 years old when, in 2004, someone took a blade to its bark, ignored the sign, and sliced the elder tree open. The trunk was gouged. Passersby wrapped duct tape around the wound, though they could not save it.

A similar fate met the Golden Spruce, more than 1,100 miles away. It had grown unobstructed for more than 300 years. The only one of its kind in the world, it rose to 165 feet and grew six feet around, seated along the Yakoun River, in the Queen Charlotte Islands of British Columbia. An activist named Grant Hadwin, then age forty-seven, took a twenty-five-inch chainsaw blade to Kiidk'yaas (meaning "ancient tree" in the Haida language), just enough to reach the core before scampering back to the mainland. When a breeze knocked the tree down, a local village of the Haida Indigenous tribe held a vigil and memorial service. The chiefs mourned the loss of an ancestor. Hadwin was arrested and charged with criminal mischief—damage in excess of 5,000 Canadian dollars—and the illegal cutting of timber.

Here may be the parable most easily related to our nation's disconnect from the arctic, from Alaska, from what we need versus what we want. Whereas visitors to the pipeline (already an eyesore) tree sought to stanch the death through duct tape, a stopgap measure, Canada went out in force to ensure it would never happen again. Makeshift solutions are the tools of the indecisive and cautiously feeble. Perhaps that difference in approach to the land and to the arctic is precisely the reason why it made sense for Hadwin, who slipped away in an expedition-grade sea kayak, promising that he would make the sixty-mile winter crossing from Prince Rupert to the courthouse in Queen Charlotte, to turn his back on the law and abscond to Alaska.

WE ARRIVE IN FAIRBANKS and it is dark again. A friend waits for me in the parking lot and we drive away. A cold snap has damaged cars across the city, freezing engine blocks that will not start, despite their built-in heaters designed to prevent that from happening. Car rental companies are all but closed during this season, with the great ice sculpture festival and its chilly slide for jumper-clad children. Frozen fog hangs along the roadways as we snake back to my apartment. Later that week, in Anchorage, I speak by phone with a researcher who focuses on the changing arctic. "Alaska is so freaking complicated," Marisol Maddox, a senior arctic analyst at the Woodrow Wilson International Center for Scholars, and an operational research analyst at PolArctic, tells me as I pace my kitchen. There is boundless frustration about what will come next for America in the arctic. So many things are restricted, secret, tense. "It was always just a place to extract things from."

Policy directed at northern territories is not hindered by a failure to see the forest for the trees but, rather, by a myopic focus on extracting what is beneath, what may kill them. The frustration felt by Maddox is a persistent part of life here, like shouting into an empty pit, only to hear no echo. Perhaps the death of the northernmost tree could be seen as a coda to the way we once viewed Alaska and the circumpolar region writ large. We have given up the fight, casting our nation into postures bent toward military clashes. It is likely, given the offshore drilling exploration occurring in the Arctic Ocean, and the continued development of ports along potential arctic sea routes, which will surely invite more tourists and more cruise ships, that developments important to national security will further fray the social and commercial fabrics of the north. Driving the Dalton Highway is conducive to this type of reflection. The loss of the tree there (at 68°01'56.1"N, 149°40'12.0"W) was the time stamp for when a global utilitarian version of stewardship at home and across the circumpolar north also died.

Not that all hope should be abandoned. A once great tree must still be born, sought, found. In July 2019, a white spruce (*Picea glauca*) was discovered in a valley of the Brooks Range, near the Canadian border. It was but a tiny, short, self-cloning spruce on a nondescript hillside. It gave rise to a flash of hope. The tree, as it takes root, will move slowly northward over the next several hundred years. Its cone will reach a hundred feet or more, resilient within the fragile, mysterious north, beneath the shadows of other trees. Unless amid our own existential or actual wars, in front of and beyond the new ice curtains we are constructing, we find some explanation to kill that tree, to kill each other, too.

Aphoristic euphemisms and symbols from my arctic dreaming stick with me. The stories, the fires, the mythical figures, and the troops seemingly everywhere. Military convoys and aircraft rumble, then fall silent, stationed elsewhere only to return. There is always a return, an unyielding tug back to a world of adumbration. Though it will be different next time, unrecognizable—as though the landscape itself were quietly remade by workers (perhaps trolls or the Neqikitsuliaq) in my absence. But not because of my absence; rather, it is because, like the mounting, unfathomable pressure of ice assaulting rock and land, stressors force change.

In Greenland, the myth goes that the Neqikitsuliaq is a gluttonous baby with giant teeth and a pronounced stomach. After its birth, Neqikitsuliaq's adoptive parents, having already fostered two other children, neglect the newest addition to their family. The other children would not grow very much otherwise, as the family has little food among them. They prioritize their healthier, older children, neglecting the Neqikitsuliaq. But secretly, the eldest daughter feeds the baby from her own food, nourishes it with water, cares for it at night. The Neqikitsuliaq grows quickly—faster than the others, overtaking them in size and might and hunger. Neqikitsuliaq changes seemingly overnight, without warning.

One night, the children are awakened by an awful noise. The Neqikitsuliaq is feasting on its mother, then its father, consuming the neglectful hands that would not feed or care for it, disemboweling the generation that first sought to save it and then left it to die. The monster spares its siblings, the next generation, for they have been kind to it. Then the Neqikitsuliaq leaves the home and spreads out across the land beyond the North Winds, vengeful and untamed, rabid, in search of other prey.

APPENDIX:
REINING IN THE ARCTIC

INTRODUCTION

We are the arctic's primary threat. We are also the arctic's biggest protector. After more than two years of discussions with arctic researchers, military leaders, and international lawmakers concerned with the future of the arctic, it has become plain to me that there are several definitive and immediate steps American lawmakers and the current presidential administration must take to not only advance our national security in the arctic but also preserve the region for generations to come.

To recap what I write about in greater detail throughout the preceding pages, the consequences of militarization and a warming planet were barely a footnote in many nations' global ambitions until about 2010 or so. Russia is leading the charge, with more military bases in the arctic, greater competency in cold-weather operations, and a fleet of icebreakers that dwarfs the maritime arctic fleets of every other nation. America and its allies have played catch-up. For the United States, this involves learning to fortify and base military units across the arctic, as well as protect critical infrastructure like oil pipelines in a climate where the country has little modern experience and capability. During a recent exercise in the north, the U.S. military was slow to recover an enemy craft because of its inexperience in the dark days of the high north and its accompanying frigid blasts. In the aftermath, the U.S. government and its military seemed to awaken to the threat of foreign dominance of the arctic—specifically Russia and China—a threat exacerbated by the changing climate. In 2022, the White House issued a National Strategy for the Arctic Region and the Department of Defense released its findings on how climate change impacts American military bases, warning of inattention to changes across the arctic. The State Department reopened a consulate in Nuuk, Greenland, and two years later, appointed an ambassador-at-large for the arctic region within the State Department, as well as a deputy assistant secretary of defense for arctic and global resilience. America's European allies, too, have been rethinking homeland security, increasing their national defense budgets (in some instances, doubling their spending) and security for critical energy infrastructure in the arctic, as they aim to boost their defense capabilities and rely less on U.S. assistance.

The American Arctic has never been more vulnerable to security issues, made worse by climate change. The fate of American-led geopolitical dominance rests along the sliver of land that encircles the seas around the North Pole. Ultimately, arctic diplomacy remains hindered by bureaucratic red tape, budget and resource misallocations, and a preoccupation with extraction of natural resources. Washington has since 2022 begun to understand the larger geopolitical whirlwind connecting the arctic to the future of the world and the nation's place within it. Quietly, beyond daily headlines and gleaned only in the circles of arctic watchers with whom I embedded and embarked for over twenty-four collective months (from Svalbard in Norway to Copenhagen, Reykjavík to Helsinki, Kodiak to Kaktovik), an extreme energy, transportation, military, and commercial conflict is consuming a region known for its disappearing albeit pristine blue glaciers, endangered wildlife, and remoteness. Now the region feels closer than ever. Climate change viewed in this context is a precursor.

To protect its sovereignty and enable its national security, Russian forces patrol the country's NSR off the coasts of Finland, Sweden, and Norway, conducting sporadic military testing, which inconveniences international fishing vessels. The same thing happens off Alaska's coast, only with American military and commercial fishing vessels. Russia has also on a number of occasions jammed GPS signals during NATO exercises in and above these waters. Meanwhile, Russia has reopened and modernized upwards of fifty Cold War–era bases along the necklace that graces its 15,000 miles of arctic coastline. The U.S. Navy has flirted with reopening its own former base in Adak, Alaska, and attempted to open a deepwater port in Utqiagvik, Alaska, but these efforts pale in comparison to what Russia and China have been able to achieve. (In one example, the two nations have jointly created a global positioning satellite network apart from international standards and regulations.)

In response to these challenges, former NATO Secretary General Jens Stoltenberg said in 2022, after Russia invaded Ukraine, that the alliance needed to strengthen its arctic defense, for the region had become an "Achilles' heel." Other arctic nations and some NATO members are catching on. In April 2023, Latvia introduced compulsory military service for all citizens regardless of gender, and it planned to build additional military bases to increase its hosting capabilities for NATO troops. Between 2022 and 2024, Sweden doubled its annual defense spending, placing a premium on space-surveillance capabilities and long- and medium-range precision weapons, typically used to defend territorial spaces. Following Stoltenberg's remarks and with neighboring increases in military spending, Denmark likewise increased its surveillance of the arctic and sought to hold more joint exercises in its northern regions. Estonia more than doubled its defense spending.

In Finland, the country's opposition to joining NATO, a territorial defense guarantor, had waned altogether. Meanwhile, in 2022, Canada pledged to spend 30 billion Canadian dollars over the coming two decades on countering Russian and Chinese military development in the arctic. Then, Canada raised its spending goal to 50 billion, to be spent by 2029.

This policy note is not aimed at the traveler on a whale-watching voyage to the high north but, rather, to those in Washington who are tasked with responsibly administering and preserving the American Arctic. It is also intended for those in charge of national defense across North America and alongside Arctic Council and NATO partners. The several steps outlined below are aimed solely at decreasing military tensions in the arctic, though opportunities abound elsewhere. Washington must be proactive in investing in the future of the arctic region—inside and outside national boundaries—by considering defense and all its related components as the primary pillars of focus for a prosperous Global Arctic.

IMPROVE OUR DEFENSE FRAMEWORK

- The military pressures described in this book directly support the need to form an Arctic Military Code of Conduct. Though the "Arctic 8" of the Arctic Council have long viewed that forum as a place apart from discussions about security as part of the original Ottawa Declaration, a code of conduct would provide the guardrails necessary to decrease the risk of miscalculation and escalation through predictability and transparency.

- Washington must address significant gaps in its defense apparatus. In 2011, the Department of Defense published a report to Congress titled "Arctic Operations and the Northwest Passage." The report identified a number of gaps to be "addressed in order to be prepared to operate in a more accessible Arctic: shortfalls in ice and weather reporting and forecasting; limitations in command, control, communications, computers, intelligence, surveillance, and reconnaissance (C4ISR) due to lack of assets and harsh environmental conditions; limited inventory of ice-capable vessels; and limited shore-based infrastructure." However, more than a decade later, these vulnerabilities still exist, along with weaknesses in various defense infrastructures, such as NORAD and the Clear Space Force Station.

- The White House should create a U.S. Arctic Intelligence Strategy and a National Intelligence Manager for the Arctic within the Office of the Director of National Intelligence to ensure cohesion across strategic communications between and originating from agencies dealing in the arctic.

APPENDIX

- Washington should restore the Aleutian Island stations, LORAN Station Attu and Adak Station. The former served as a navigational aid for mariners in the North Pacific until it closed in 2010, and the latter was in use until 1997, when it was repurposed as a civilian air facility. According to Alaska Senator Sullivan, investing in Adak Station's military infrastructure is crucial for force projection, especially now when tensions in the Pacific are high: There have been "dozens of incidents near Alaska involving Russian and Chinese military assets." Additionally, both Adak Station and LORAN Station Attu should be modernized for sea monitoring and ISR to demonstrate force projection and support the monitoring operations currently handled by a single USCG vessel.

- Washington should engage with allies to replace naval contractors to streamline the National Security Cutter and Polar Security Cutter programs, ensuring the timely delivery of vessels toward not only arctic security and force projection but also the coast guard's larger mandates of global drug interdiction and scientific research. Although the One Big Beautiful Bill Act pledged a significant $9 billion for cutter and ice-capable vessel development, our shipbuilding infrastructure remains inadequate and must be strengthened.

- "Arctic nations' search-and-rescue (SAR) capabilities vary widely, and none is sufficiently equipped to meet the needs of commercial operators when emergencies inevitably arise in this rapidly developing region," according to a 2023 report titled, "A Dose of Reality: Search and Rescue for Arctic Sustainable Development." Currently, search-and-rescue missions in the arctic are often hampered by a declared capacity rule, which states that there are only a certain number of people safely allowed aboard a vessel. Washington should adapt this law to explicitly allow for the boarding of all stranded crew members and passengers in rescue missions, and more broadly speaking, urge the Arctic Council to "charter a committee with regular meetings and exercises and [develop] a common operating procedure database for registering all national, local, and private SAR assets available to support regional operators."

- Washington must redistribute its arctic assets to ideal locations. Fifth-generation fighters are placed in Alaska, and there currently exist arctic Airlift Wing groups in New York. It's only natural that similar assets are stationed at the Pituffik Space Base and positioned where they could readily support Danish and Greenlandic partners.

- As climate change ravages the planet, the arctic region is disproportionately affected. Fluctuating environmental conditions affect the utility of valuable maritime routes, thus impacting our national security. In February 2025, President Trump fired everyone from a NSF program called Office of the Chief of Research Security Strategy and Policy, which aimed at preventing foreign adversaries from accessing proprietary and governmental scientific research. Washington must direct the NSF both to make this office whole again and to continue to carry out its important research in cold-weather infrastructure and habitats.

- The White House must follow through on its commitment to decarbonize its economy and achieve net zero by 2050, for the safety of the region. By advocating for carbon neutrality, the United States, as an Arctic nation, could lead the charge to enhance global security.

- Washington should advocate for the "Arctic Security Initiative" to fund public–private infrastructure projects to increase awareness and safety measures. Because the private sector handles most submarine cables and pipelines, strengthening cooperation between the public and private sectors would "deter threats to critical infrastructure."

- By ratifying the United Nations Convention on the Law of the Sea, Washington could keep pace with "Russia and other nations staking claims to resources extending to the North Pole." Although this may be a dangerous proposition for both the outstanding Beaufort claim and America's own disagreement over Canada's sovereign control of the

Northwest Passage, ratifying the law would be a diplomatic maneuver to offer an olive branch to an ally, strengthening ties in good faith while standing strong against a resurgent arctic tyrant.

STRENGTHENING TIES WITH ALLIES IN EUROPE AND GREENLAND

- If we want to maintain and strengthen our open communication lines with allies and bolster collaborative defense capabilities, we would be wise to appoint a new Arctic Ambassador-at-Large.

- To assess threats in a more comprehensive manner, the White House should implement a long-term program of intelligence community civilian training and integration with arctic allies and partners. Washington would "benefit from a pan-Arctic approach that views the region as a whole, instead of through the current, fragmented lens split across three different combatant commands (European Command, Northern Command, and Indo-Pacific Command). Intelligence strategies are important for integrating the vast intelligence community with key partners within the larger interagency, with state, local, and tribal entities; the private sector, and U.S. allies to encourage cooperation, coordination, and mutual support."

- Working collaboratively with allies is imperative, especially within today's geopolitical context: Russia's invasion of and continuous assault on Ukraine, Iran's blind support of Russia's activities, China's and Russia's broadening influence on the Middle East and Global South, and a gradual shift away from Western ideology. By sharing intelligence and developing "contingency plans to detect, deter and respond to hybrid and 'grey-zone' activities," Washington and its allies can set up guardrails to avoid escalation.

- One such avenue for securing our allies' autonomy and safety is in supporting Greenland's bid for both EU membership and acceptance into the European Free Trade Association (EFTA), while abandoning all attempts to purchase it. If Washington doesn't demonstrate unyielding support for Greenland's autonomy—which we have yet to witness in Trump's second term—we can expect Nuuk to turn to Russia and China, especially because Greenlanders are more concerned with their domestic politics and obtaining autonomy than with geopolitical competition.

- Washington would be wise to continue its support of geological surveys using geospatial imagery for exploring the mineral and rare-earth wealth across Greenland. Doing so would empower the country to stand on its own financially, which would align with American interests across its material wealth and political agendas.

- Washington should back climate initiatives that support the local and national economies in Greenland, such as Minik Rosing's "rock flour" project. Bolstering homegrown alternatives to foreign investment will ensure U.S. influence in the region and counter foreign investment seeking to destabilize the region and drive wedges between Denmark, Greenland, and its allied partners. Supporting this project in particular would help Alaskans who are turning to agricultural initiatives as aired land becomes a reality across the thawing state.

Further reading on the arctic and the changes occurring there can be found in the Selected Bibliography section in the back of this book and at https://kennethrrosen.com/arctic.

ACKNOWLEDGMENTS

THIS BOOK WOULD NOT have been possible without:

Elizabeth Ralph, at *Politico* magazine, for sending me on assignment to Svalbard. At Javelin, my agents Keith Urbahn and Matt Carlini. At Simon & Schuster, Ian Straus, Priscilla Painton, Robert Messenger, and the copy editors, legal reviewers, and designers. For her early encouragement and enduring support, Sarah Fallon. MacDowell and the Jan Michalski Foundation both provided time and material support; the Snedden Foundation in Fairbanks and its chair Virginia Farmier; and the de Groot Foundation, for their financial support.

Paul Roszkowski, the entire U.S. Coast Guard Motion Picture, Television and Author Liaison Office, District 17 and the Kodiak Air Station, Anthony L. Russell from the Center for Arctic Study and Policy, and in particular the deck force of the U.S.C.G.C. *Stratton*, Lieutenant (junior grade) Claire Sullivan, Boatswain's Mate Second Class Stephen Davis, and Dedicated Crew Chief Anthony Bemis.

The staffs, public affairs liaisons, FOIA officers, and military personnel of the Försvarsmakten, Fosvaret, Finnish Border Guards, the army's 11th Airborne Division, Joint Base Elmendorf-Richardson, Eielson Air Force Base, Fort Wainwright, the Army Engineer Research and Development Center, and the Ted Stevens Center.

Chad, Haley, Abby, Ben, Alex, Kevin, and the entire staff working to support and maintain the Toolik Field Station year-round. The staff at the Institute of Arctic Biology at the University of Alaska Fairbanks for their assistance and support. Pamela Miller and her Arctic Conservation Library, Terry Callaghan and the INTERACT network, and the countless researchers and scientists who patiently explained to me their life's work and its implication for our planet's future.

Dr. Michael E. Lynch and Dr. Michele Devlin at the U.S. Army War College; Dr. Ryan Burke at the U.S. Air Force Academy; Commander Thomas Gjesdal of the Royal Norwegian Navy; Lieutenant Col. Semming Rusten and Dr. Lon Strauss at the U.S. Marine Corps University; Ambassador Mike Sfraga; Heather Conley; Sherri Goodman; the U.S. diplomatic mission staffs in Reykjavík, Tallin, Oslo, Tromsø, Nuuk, Vilnius, the European Union, and NATO, for their assistance in providing interviews, invaluable insight, and sharing in my interests in the changing arctic.

Greenland's Arctic Hub, especially Ole Ellekrog, the Greenland Institute of Natural Resources; and Dr. Maria Ackrén, Rikke Østergaard and Dr. Javier Arnaut at Ilisimatusarfik/University of Greenland.

Michael Dulaney, Dr. Marc Lanteigne, Dr. Rob Huebert, Dr. Adam Lajeunesse, Dr. Russell Potter, Dr. Michael Wenger, Dr. Gunnar Rekvig, Vadim Puzyrev (Zhigansky), Robert Meissner, Caecilie Christensen, Toby Warsitha, and Lisa Jacobsson for translations and transcriptions, peer reviews, and book lists. Eric Feely and Rebecca Redlich, for research and fact-checking assistance.

Arnfinn Morsund Sjøenden; my friends on Svalbard, Marte, Elizabeth, Natalia, Masha, Timofey, and Holt; Inger-Adele Caplan, Elizabeth Bourne, Ellena Savage, Natalia Maksimishina, David Valicenti, Lucas Rencoret, Andy Glas, Astrid Fadnes, Thomas Nielson, Tore Rørbæk, Holt Hancock, Abraham Steiner, Jon Sadler, Hrafnkell Gissurarson, Claudio Revere, Amy Loeffler, Kara Stanley, Tasket, Deeplow, Rocodes, and the Freedom of the Press Foundation for technical, emotional, and spiritual support.

An exhaustive list would make a dense second volume. For those I have regrettably and certainly overlooked, or who have not appeared elsewhere

ACKNOWLEDGMENTS

in the book, know that what you taught me can be found behind each and every sentence.

Lastly, a line on a page in a book on a dusty shelf cannot return the time I lost with my family while traipsing above and around the Arctic Circle. Elettra, anywhere you are is home. This book is as much yours as mine. *Crederò sempre in noi.* And to our children, who gingerly embraced harsh, frigid temperatures at home and abroad, relished in Icelandic and Nordic folktales at bedtime, and held fast to my many postcards depicting fire and ice: though when you are grown, there will be little that remains of the arctic and northern wildernesses as I've come to know them, I nevertheless wish that, as my arctic dreaming has been for me, you allow your lives to grow into vast, wild spaces, and learn that through them you may avail yourselves of anything.

NOTES

DRAWING ON INTERVIEWS WITH more than 400 individuals presenting a snapshot of the circumpolar north, including representatives of Arctic Council observer states and Indigenous communities, NATO military
planners and foreign diplomats, reports and U.S. Senate legislation from the
last decade, government findings published by the five Arctic Ocean littoral
states, arctic strategies released throughout the last decade by more than
twelve nations, and government and military agency documents received
through dozens of Freedom of Information Act requests (https://kennethr
rosen.com/arctic), this book contends that America and its allies are engaged
in a new Cold War in the arctic regions. Whereas its allies have taken necessary steps to confront the challenges facing those regions, America has not
gracefully risen to meet the occasion. A once-strong arctic power has fallen.

To witness those preparations and failings, to observe those efforts to
regain America's place on the arctic world stage, I primarily traveled above
the Arctic Circle on excursions over twenty-four months, beginning in September 2022 by embarking with coast guard and fishing vessels, warships
and cruise liners, icebreakers and Zodiacs, military aircraft and commercial transports, and on snow machines, electric cars, and electric scooters.
I attended the 2023 NATO summit in Vilnius, Lithuania, at which Finland
formally joined NATO, followed soon after by Sweden; I audited classes at
the U.S. Marine Corps University, endured a sterile and bureaucratic meet-

ing for icebreaker mission planning in Alexandria, Virginia; received my Merchant Mariner Credentials from the U.S. Coast Guard; and became a certified Wilderness First Responder in Iceland.

During those years, I lived and spoke with those who have experienced the consequences and the impacts of these rapidly shifting north lands. Across the circumpolar north, I befriended those who lived apart from the houses of power, wandered in their mountains, ate in their weather-beaten cabins under feet of snow, and was aboard their vessels gliding through ice, as well as taking part in their celebrations beneath the northern lights, witnessing the arrests of their outlaws, and wading over endless murky tundra. I celebrated Rosh Hashanah with the last Jews of Finnmark in the back of a lingerie shop; was accosted twice in Icelandic bars while living for a time out of a recreational vehicle; participated in close-quarter combat drills on an ad hoc shooting range amid the Arctic Ocean ice sheet; lectured in Fairbanks; reunited with old friends in Anchorage; boarded with soldiers and sailors, scientists and seclusionists.

A few clarifications: For the sake of narrative fluidity, I have used quotes when those words were spoken directly to me and recorded, and I took the liberty of using quotes as relayed to me by people who were recalling what they had once said and heard. However, I have corroborated secondhand accounts with no fewer than two additional sources, sometimes supplementing personal recollection with publicly available documents. Most interviews with U.S. and foreign government officials were conducted off the record or on deep background, except where noted through direct attribution. I've made caustic and provocative choices to name certain places in certain ways, to lowercase the word "arctic," if only as a small protest against modern colonialism.

Inevitably, the scope for such a book as this is immense, and not all issues have been given equal weight. Entire libraries have been written about subjects that appear only as marginalia here. I make no claim to be an authority on the arctic or its environs; I do not speak the local languages; I own little stake in the region, both personally and professionally; and the only formal studies I undertook of the region were impersonal and brisk. The following is a more detailed record of those efforts.

NOTES

A NOTE TO READERS

1   *campaign to "get" and secure Greenland:* Michael Crowley and Maggie Haberman, "Inside Trump's Plan to 'Get' Greenland: Persuasion, Not Invasion," *New York Times,* April 10, 2025, https://www.nytimes.com/2025/04/10/us /politics/trump-greenland-denmark.html.

FLAGS

3   *manual to computer-assisted navigation:* Adding to the general mayhem surrounding the definition of the arctic, there are three northern poles, with varying degrees of utility and importance: the geographic North Pole, at 90 degrees north; the North Magnetic Pole; and the North Geomagnetic Pole.

4   *glaciers in various phases of collapse:* Jon Gertner, "Can $500 Million Save This Glacier?," *New York Times Magazine,* January 6, 2024, https://www.ny times.com/2024/01/06/magazine/glacier-engineering-sea-level-rise.html.

4   *polar expedition of Charles Francis Hall in 1871:* The Smithsonian National Museum of American History, Washington, D.C., has a phenomenal collection of documents from Hall's arctic exploration; see the Charles Francis Hall Collection.

4   *slightly longer than a half mile:* Hans Island is 1.3 square kilometers, or 321 acres.

4   *tassel of water:* Hans Island is located at 80.8269 degrees north, 66.4597 degrees west.

4   *American explorer Elisha Kent Kane:* For more on his expeditions, see Elisha Kent Kane and John Franklin, *Arctic Explorations: The Second Grinnell Expedition in Search of Sir John Franklin, 1853, '54, '55* (Childs & Peterson, 1856), https://www.loc.gov/item/04019331.

5   *was officially drawn:* Elin Hofverberg, "The Hans Island 'Peace' Agreement Between Canada, Denmark, and Greenland," Library of Congress, blog, June 22, 2022, https://blogs.loc.gov/law/2022/06/the-hans-island-peace-agree ment-between-canada-denmark-and-greenland.

NOTES

5   *Dome Petroleum arrived on:* "Taissumani, August 29, 1871: Hall Names Hans Island," *Nunatsiaq News,* August 26, 2005, https://nunatsiaq.com/stories/article/taissumani_august_29_1871_-_hall_names_hans_island.

5   *"Welcome to the Danish Island":* Dan Levin, "Canada and Denmark Fight over Island with Whisky and Schnapps," *New York Times,* November 7, 2016, https://www.nytimes.com/2016/11/08/world/what-in-the-world/canada-denmark-hans-island-whisky-schnapps.html.

6   *politics ensued:* "Kingdom of Denmark Strategy for the Arctic 2011–2020," government report, August 2011, https://uniset.ca/microstates/mss-denmark_en.pdf.

6   *settling such territorial or inter-governmental disputes:* "Why It Took 50 Years to Resolve Canada and Denmark's Dispute over a Tiny Arctic Island," *National Geographic,* June 27, 2024, https://www.nationalgeographic.com/history/article/how-hans-island-sparked-whisky-war-between-canada-denmark.

6   *The final agreement:* Peter Beaumont, "Canada and Denmark End Decades-Long Dispute over Barren Rock in Arctic," *The Guardian,* June 14, 2022, https://www.theguardian.com/world/2022/jun/14/canada-denmark-end-decades-long-dispute-barren-rock-arctic-hans-island.

7   *"What happens here impacts everywhere":* Rear Admiral Rune Andersen (commander of the Norwegian Navy), telephone interviews with author, 2022 and 2023.

7   *Every year the Arctic Ocean:* The data are from the NOAA, which regularly offers arctic "report cards" on everything from the extent of sea ice to predicted storm systems. See "2023 Arctic Report Card: Image Highlights," NOAA Climate.gov, report, December 12, 2023, http://www.climate.gov/news-features/understanding-climate/2023-arctic-report-card-image-highlights.

7   *Greenland's ice cap is another story:* Elizabeth Kolbert, "When the Arctic Melts," *The New Yorker,* October 7, 2024, https://www.newyorker.com/magazine/2024/10/14/when-the-arctic-melts.

7   *thermohaline circulation:* "Thermohaline Circulation: Currents Tutorial," National Oceanic and Atmospheric Administration, n.d., https://oceanservice.noaa.gov/education/tutorial_currents/05conveyor1.html.

8   *Marine life is being displaced:* "Biodiversity and Nature," World Wildlife
     Fund: Arctic, information sheet, n.d., https://www.arcticwwf.org/our-prior
     ities/biodiversity-and-nature.

8   *"zone of peace":* From Mikhail Gorbachev's Murmansk speech, more fully
     recounted in succeeding chapters and cited in the first note for page ten below.

9   *nuclear submarine collided with:* John H. Cushman, "Two Subs Collide Off
     Russian Port," *New York Times,* February 19, 1992, https://www.nytimes
     .com/1992/02/19/world/two-subs-collide-off-russian-port.html.

9   *their frequency and their consequences:* Gunhild Hoogensen Gjørv, "Security
     and Geopolitics in the Arctic: The Increase of Hybrid Threat Activities in the
     Norwegian High North," Hybrid CoE Working Paper 30, March 27, 2024,
     https://www.hybridcoe.fi/publications/hybrid-coe-working-paper-30-secu
     rity-and-geopolitics-in-the-arctic-the-increase-of-hybrid-threat-activities
     -in-the-norwegian-high-north.

9   *taking great interest in:* Of the many arctic collections and archives, I found
     two to be particularly useful in realizing the once-prominent American—
     indeed, global—interest in the arctic. See the Alaska and Polar Regions Col-
     lections and Archives at the University of Alaska Fairbanks (https://library
     .uaf.edu/aprca) and the various arctic collections at the University of Illinois
     Urbana-Champaign (https://library.illinois.edu/rbx/research-instruction/re
     search-guides/arctic-collections).

10   *Gorbachev said in 1987:* The full remarks of Gorbachev's speech and the
     Murmansk Initiative on October 1, 1987, are available at Internet Archive,
     https://web.archive.org/web/20241127164607/https://www.barentsinfo.fi
     /docs/Gorbachev_speech.pdf.

10   *arctic may be divided by latitude into:* I decided not to capitalize *arctic* unless
     it is used in proper names, for there is no one definition of the arctic. See
     Changes in the Arctic: Background and Issues for Congress, report, January
     2024. This includes a citation for the article I wrote for *Politico* in Decem-
     ber 2022, which was the genesis for this book. It reads, in part, "There are
     multiple definitions of the Arctic that result in differing descriptions of the
     land and sea areas encompassed by the term. Policy discussions of the Arctic

can employ varying definitions of the region, and readers should bear in mind that the definition used in one discussion may differ from that used in another."

10   *population 13 million:* Ilmari Hustich, "The Population of the Arctic, Sub-arctic, and Boreal Regions," *Polar Geography* 3, no. 1 (1979): 40–48, https://doi.org/10.1080/10889377909377100; Timothy Heleniak, "The Future of the Arctic Populations," *Polar Geography* 44, no. 2 (2020): 136–52, https://doi.org/10.1080/1088937X.2019.1707316.

10   *resulting in definitions as varied:* Attempting a breakdown of definitions for the regions is simply bonkers. There are many delineating lines drawn to determine what constitutes the arctic, and as I hope the book makes clear, even these are subject to change, as follows:

-   The 10°C July isotherm line, defined as the area where the average temperature for the warmest month (July) is below 10°C/50°F.
-   The line in the *Arctic Human Development Report* (*AHDR*).
-   The line drawn by the Arctic Emergency Prevention, Preparedness and Response (EPPR).
-   The line drawn by Conservation of Arctic Flora and Fauna (CAFF).
-   The line drawn by the Arctic Monitoring and Assessment Programme (AMAP).
-   The Arctic tree-line boundary that is at the northernmost latitude in the Northern Hemisphere where trees can grow; farther north, it is too cold year-round to sustain trees.

See "Arctic Definitions," Arctic Portal.org, information sheet, n.d., https://arcticportal.org/education/quick-facts/the-arctic/3448-arctic-definitions.

11   *Smaller-growth trees creak:* A fascinating tour of the boreal forests around the world is offered in Ben Rawlence, *The Treeline: The Last Forest and the Future of Life on Earth* (St. Martin's, 2022).

11   *minus 80 degrees are typical:* Ian Livingston, "Siberia Sees Coldest Air in Two Decades as Temperature Dips to Minus-80," *Washington Post,* January 11, 2023, https://www.washingtonpost.com/weather/2023/01/11/siberia-russia-extreme-cold.

NOTES

11   *Defense Threat Reduction Agency's Strategic Forum:* A recording of Dr. Brandenberger's presentation and the accompanying PowerPoint slide deck were obtained through a Freedom of Information Act request. Defense Threat Reduction Agency, FOIA Case No. 23-128, submitted August 28, 2023. For full documents and recording, go to https://kennethrrosen.com/arctic.

12   *known as horizontal escalation:* There are many great academic papers that go into this process and subsequent strategies to detect, mitigate, and deter escalation. One in particular, by Michael Fitzsimmons from the Air University, illuminates the processes as a means for confronting Russia. See "Horizontal Escalation: An Asymmetric Approach to Russian Aggression?," *Strategic Studies Quarterly* 13, no. 1 (Spring 2019): 95–133. For the formal definition, see "horizontal escalation," *Cambridge Dictionary*, https://diction ary.cambridge.org/us/dictionary/english/horizontal-escalation.

12   *more frequent discussions in Washington about:* These tabletop exercises were relayed to me by two former White House officials who worked on the exercises and participated in similar events as recently as 2023, though undoubtedly they continue to this day.

12   *As Mark Rutte:* Lulu Garcia-Navarro, "The Interview: The Head of NATO Thinks President Trump 'Deserves All the Praise,'" *New York Times,* July 5, 2025, https://www.nytimes.com/2025/07/05/magazine/mark-rutte -interview.html.

12   *deploy forces quickly:* Hilde Gunn Bye and Birgitte Annie Hanson, "US Arctic Paratroopers Practiced Rapid Deployment in Northern Norway," *High North News,* March 20, 2024, https://www.highnorthnews.com/en/us-arc tic-paratroopers-practiced-rapid-deployment-northern-norway. Traversing the North Pole by air is no less difficult than by icebreaker. Finnair offers flights to Asia that transit the North Pole and must be scheduled to avoid solar flare activity. For completing the journey, passengers receive a certificate once they fly over the pole.

12   *two twenty-six-foot-long Russian submersibles:* Alex Shoumatoff, "The Arctic Oil Rush," *Vanity Fair,* April 14, 2008, https://www.vanityfair.com/news /2008/05/arctic_oil200805.

13   *"This isn't the fifteenth century"*: "Canada Mocks Russia's '15th Century' Arc-
      tic Claim," Reuters, August 9, 2007, https://www.reuters.com/article/world
      /canada-mocks-russias-15th-century-arctic-claim-idUSN02464985.

13   *a* Time *Person of the Year*: "Person of the Year 2007," *Time*, December 19, 2007,
      https://content.time.com/time/specials/2007/personoftheyear/0,2875
      7,1690753,00.html.

13   *"Everything will be all right"*: Shoumatoff, "Arctic Oil Rush."

13   *upgrades a base on*: "Inside the Military Base at the Heart of Putin's Arc-
      tic Ambitions," CNN, April 4, 2019, https://edition.cnn.com/2019/04/04
      /europe/russia-arctic-kotelny-island-military-base/index.html.

13   *United States relocates fifth-generation fighter*: Amanda Miller, "F-35 Squadrons in
      Alaska Shift to Full Operations as 'Advanced Threats' Grow 'More Lethal,'" *Air &
      Space Forces,* August 10, 2022, https://www.airandspaceforces.com/f-35-squad
      rons-in-alaska-shift-to-full-operations-as-advanced-threats-grow-more-lethal.

13   *research balloon soaring over the*: Stephanie Hogan, "When a Suspected
      Chinese Spy Balloon Flew over Canada, Why Didn't We Shoot It Down?,"
      CBC News (Canada), February 9, 2023, https://www.cbc.ca/news/canada
      /spy-balloon-canada-norad-questions-1.6742695.

13   *nation sends two*: Lt. Michelle Pelissero, "U.S. Navy Kicks Off ICEX 2020 in
      Arctic Ocean," U.S. Northern Command, March 6, 2020, https://web.archive
      .org/web/20241206141209/https://www.northcom.mil/Newsroom/News
      /Article/Article/2104690/us-navy-kicks-off-icex-2020-in-arctic-ocean.

13   *another nation sends three*: "Комплексная Арктическая Экспедиция ВМФ
      России и РГО «Умка-21»," YouTube, March 26, 2021, https://www.youtube
      .com/watch?v=CkXbyklunp4.

13   *this time replaced by*: Atle Staalesen, "In Fascist-Inspired Crusade, Warriors
      from Moscow's War of Aggression Wave Z-Flags on North Pole," *Barents
      Observer,* June 12, 2024, https://www.thebarentsobserver.com/arctic/in-fas
      cistinspired-crusade-warriors-from-moscows-war-of-aggression-wave-zflags
      -on-north-pole/102140.

14   *the kidnapping and displacement of thousands*: "Russia's War on Ukraine:
      Forcibly Displaced Ukrainian Children," Think Tank (European Parliament),

February 11, 2025, https://www.europarl.europa.eu/thinktank/en/document/EPRS_BRI(2023)747093.

14   *perhaps most of all in America:* To conduct this informal, qualitative survey, I interviewed professors and department heads at the five military academies and graduate-level institutions for the military.

14   *One resident of the European Arctic:* This source requested anonymity because of their complex residency status and the tight-knit community in which they live.

CLOUDBERRIES

15   *as it was called in Greek:* There are plenty of books and articles discussing the Hyperboreans and their land, but perhaps of most interest is Grace Harriet Macurdy, "The Hyperboreans," *Classical Review* 30, no. 7 (1916): 180–83, https://doi.org/10.1017/S0009840X0001060X. Hyperborea has become a fixture of academic papers discussing Russia's history of High North exploration and domineering.

16   *"though most probably":* Oxford University is home to the Scott Polar Research Center and the publisher of some of the finest material on the polar regions. The information in this and the succeeding paragraph is derived from a variety of sources, though the impetus for most of it was Sir John Richardson, *The Polar Regions* (Cambridge University Press, 2014).

16   *"the limits of the unknown":* Fridtjof Nansen, *Farthest North,* 2 vols. (Harper & Brothers, 1861).

17   *the meaning of "sea level":* Brooke Jarvis, "Our Very Strange Search for 'Sea Level,'" *The New Yorker,* August 19, 2024, https://www.newyorker.com/magazine/2024/08/26/our-very-strange-search-for-sea-level.

17   *euphoric tales of nineteenth-century:* See this book's Selected Bibliography, as it's a great place for the curious and the voracious to start.

17   *has been documented since:* One well-cited academic work that has helped frame debates about the heightened geopolitical tensions in the arctic since the early aughts—often associated with "new Cold War" narratives—is Michael

Byers, *Who Owns the Arctic? Understanding Sovereignty Disputes in the North* (Douglas & McIntyre, 2010).

17 *was slow to recover an enemy craft:* Aside from a bevy of more recent incidents, the U.S. military has a long history of difficulties operating in the cold. See Lt. Col. M. G. Coe, "Cold Winter 83, Norwegian Army OJT, After Action Report," 1983, Exercises, box 166, folder 6, Archives, Marine Corps History Division, 2; John Vinocur, "U.S. Marines Struggle to Cope with Norway's Arctic," *New York Times,* May 26, 1979; Mats R. Berdal, *The United States, Norway, and the Cold War, 1954–60* (St. Martin's, 1997), 48, 174–75; and David B. Crist, "A New Cold War: U.S. Marines in Norway and the Search for a New Mission in NATO," in *New Interpretations in Naval History: Selected Papers from the Fourteenth Naval History Symposium Held at Annapolis, Maryland, 23–25 September 1999,* ed. Randy Carol Balano and Craig L. Symonds (Naval Institute Press, 2001), 344–45.

18 *front door key to a balloon:* "Alaskan Hotel & Bar History," The Alaskan Hotel & Bar, n.d., https://www.thealaskanhotel.com/history.

18 *"There's no horizon":* Unnamed man, told to author, 2013.

20 *until the summer of 2023:* Thomas Nilsen, "Climate Change Brings Cloudberry to Svalbard," *Barents Observer,* October 3, 2023, https://thebarentsobserver.com/en/arctic/2023/10/climate-change-brings-cloudberry-svalbard.

20 *I am fearful of crushing:* It is eventually a worker at Poste Nord who crushes them.

OUT THE ROAD

22 *a decidedly paradoxical island:* Russia prefers to reference "Spitsbergen" when discussing the broader cluster of islands, rather than "Svalbard," in part because it wishes not to recognize Norway's sovereignty over the archipelago.

22 *Dutch explorer Willem Barentsz:* Gerrit de Veer, *The Three Voyages of William Barentsz to the Arctic Regions: 1594, 1595 and 1596,* ed. Laurens Koolemans Beynen (Hakluyt Society, 1853).

22 *Fifteen nations:* Though widely referenced today as the "Svalbard Treaty," the original treaty was titled "Spitsbergen Treaty" (https://www.spri.cam.ac.uk/resources/infosheets/spitbergentreaty.pdf) or "Traite relatif a l'Archi-

pel du Spitsberg" (https://treaties.un.org/doc/Publication/UNTS/LON/Volume%202/v2.pdf).

22   *is home to around:* Population of Svalbard, Statistics Norway, September 26, 2024, https://www.ssb.no/en/befolkning/folketall/statistikk/befolkningen-pa-svalbard.

22   *polar bears live:* Sarah Kuta, "Meet the Women Protecting Travelers from Polar Bears in Svalbard," *Condé Nast Traveler,* November 17, 2022, https://www.cntraveler.com/story/meet-the-women-protecting-travelers-from-polar-bears-in-svalbard.

22   *the best of humanity:* Jenny Skagestad, in a TEDx Talk titled "Svalbard—Canary in the Coal Mine Goes Green," referred to Svalbard as a "miniature Norway, and Norway is a miniature of the world." A logical extension of this is that Svalbard is the world, and this is true in many ways, perhaps most strikingly in its representation of the world as it was designed to be, the hopeful constructs formed after a war to end all wars. See Zdenka Sokolíčkovà, *The Paradox of Svalbard: Climate Change and Globalisation in the Arctic* (Pluto Press, 2023), as a rather enriching view of the archipelago, its communities, histories, ambitions, and quandaries.

23   *videos to its website:* See "How to Dress in Svalbard," Visit Svalbard, n.d., https://en.visitsvalbard.com/visitor-information/travel-information/how-to-dress-in-svalbard.

23   *ice would have formed:* "Losing Ice in Svalbard," NASA Earth Observatory, August 19, 2017, https://earthobservatory.nasa.gov/images/92325/losing-ice-in-svalbard.

23   *bringing out cloudberries and bringing in more people:* "The $10,000 Cruise That Recently Became Possible—and Might Not Last for Long." *Washington Post,* October 27, 2024, https://www.washingtonpost.com/climate-environment/2024/10/27/norway-svalbard-cruise-industry-tourism-global-warming.

24   *real glacier ice:* Two stories bracket nearly a decade of coverage about such curiosities, constituting the majority of coverage about the region. See Ole Ellekrog, "Greenland Startup Begins Shipping Glacier Ice to Cocktail Bars in the UAE," *The Guardian,* January 9, 2024, https://www.theguardian.com/world/2024/jan/09/greenland-startup-shipping-glacier-ice-cocktail-bars-uae-arctic-ice; Mary Catherine O'Connor, "A Norwegian Company's Plan

to Make Ice Cubes out of Glaciers Unsettles Some," *The Guardian,* April 4, 2015, https://www.theguardian.com/vital-signs/2015/apr/04/svaice-ice-cubes-glacier-melt-cocktails-bars.

24   *"out the road":* Years ago, I published a personal essay about this phenomenon; see Kenneth R. Rosen, "Out the Road," *The Rumpus,* July 5, 2016, https://therumpus.net/2016/07/05/out-the-road.

24   *"ready for Iceland's wonders!":* I visited Iceland on no fewer than three extended trips over two years, and on occasions when I did take the bus, this never failed to happen.

25   *old miners' quarters:* See "About Us," owner's webpage, Polarriggen, 2025, https://polarriggen.com/about-us.

25   *parachuting onto the North Pole:* Trude Pettersen, "Russian Military Instructors Plan to Land on Svalbard," *Barents Observer,* April 7, 2016, https://www.thebarentsobserver.com/security/russian-military-instructors-plan-to-land-on-svalbard/152833.

25   *suggested invading the island:* Denis Zagore and Thomas Nilsen, "Halfwit Lawmaker Says Russia Should Take over Svalbard," *Barents Observer,* January 15, 2025, https://www.thebarentsobserver.com/news/halfwit-lawmaker-says-russia-should-take-over-svalbard/423172.

25   *investment opportunities in newer facilities:* Thomas Nilsen, "Isolated Russia Invites Faraway Countries to Upcoming Svalbard Science Center in Pyramiden," *Barents Observer,* October 30, 2023, https://thebarentsobserver.com/en/arctic/2023/10/ghost-town-pyramiden-will-be-home-russias-planned-international-svalbard-science.

25   *Russian drone operators:* This has happened on a number of occasions since 2022. Some of those are listed here. See Astri Edvardsen, "Russian Citizen Arrested in Northern Norway for Drone Flights on Svalbard," *High North News,* October 20, 2023, https://www.highnorthnews.com/en/russian-citizen-arrested-northern-norway-drone-flights-svalbard; "Norway Arrests Ship's Officer for Flying Drone as It Tightens Port Security," *Maritime Executive,* October 7, 2024, https://maritime-executive.com/article/norway-arrests-ships-officer-for-flying-drone-as-it-tightens-port-security.

26   *some were paid handsomely*: Elizaveta Vereykina, "Son of Former Putin's Ally Awarded 2,7 Million Kroner Compensation," *Barents Observer*, October 15, 2024, https://www.thebarentsobserver.com/news/son-of-former-putins-ally-awarded-27-million-kroner-compensationnbsp/326634.

26   *voting and driving privileges revoked*: Kenneth R. Rosen, "A Shadow over Svalbard," *The Dial*, May 7, 2024, https://www.thedial.world/issue-16/svalbard-international-rights-norway.

26   *for military research*: Didi Kirsten Tatlow, "China's Expanding Arctic Ambitions Challenge the U.S. and NATO," *Newsweek*, July 21, 2024, https://www.newsweek.com/2024/08/09/china-russia-us-arctic-north-pole-strategy-svalbard-norway-sea-route-1916641.html.

26   *rallies, flotillas, and paraded*: Thomas Nilsen, "Russian Diplomat Staged Navy Parade at Norway's Svalbard Archipelago," *Barents Observer*, July 21, 2023, https://thebarentsobserver.com/en/security/2023/07/russian-consul-staged-navy-parade-norways-svalbard-archipelago.

26   *submarine cable connecting the islands*: Benjamin Fredriksen, Beth Mørch Pettersen, Gyda Katrine Hesla, Inghild Eriksen, and Håvard Gulldahl, "Russiske Trålere Krysset Kabler i Vesterålen Og Svalbard Før Brudd—Nordland," *NRK*, June 26, 2022, https://www.nrk.no/nordland/xl/russiske-tralere-krysset-kabler-i-vesteralen-og-svalbard-for-brudd-1.16007084.

26   *Fause has his suspicions*: Lars Fause (governor of Svalbard), interview with author, October 2022, Longyearbyen, Svalbard, Norway.

28   *a group of residents*: Rolf Stange, "Citizens Without Voting Rights: Press Release of the 'Unwanted Foreigners,'" Spitsbergen-Svalbard, news release, October 9, 2023, https://www.spitsbergen-svalbard.com/2023/10/09/citizens-without-voting-rights-press-release-of-the-unwanted-foreigners.html.

28   *a fixture of proposals*: Georgi Kantchev, "Russia and China Defy the West Deep in the Arctic," *Wall Street Journal*, September 28, 2024, https://www.wsj.com/world/svalbard-russia-china-arctic-trade-e6187bd8.

28   *chartering its own boats*: Thomas Nilsen, "Russia Says It Will Start Direct Tourist Voyages to Svalbard," *Barents Observer*, April 3, 2024, https://www

.thebarentsobserver.com/arctic/russia-says-it-will-start-direct-tourist-voyages-to-svalbard/122129.

28 *would build a research station:* "Minister: Russia Will Develop International Scientific Station on Spitsbergen," *TASS* (Russian News Agency), April 6, 2023, https://tass.com/economy/1600365.

28 *signed a cooperation agreement:* Elisabeth Braw, "Arctic Harmony Is Falling Apart," *Foreign Policy,* May 15, 2023, https://foreignpolicy.com/2023/05/15/russia-china-arctic-cooperation-svalbard.

28 *"combat terrorism, illegal migration":* Sergey Shiryaev, "A Memorandum of Understanding Was Signed Between the Federal Security Service of the Russian Federation and the Coast Guard of the People's Republic of China," State Television and Radio Broadcasting Company "Murman," April 24, 2023, https://murman.tv/news-n-11036—podpisan-memorandum-o-vzaimoponimanii-mezhdu-federalnoj-sluzhboj-bezopasnosti-fr-beregovoj-ohranoj-knr.

29 *security and prosperity:* Various Norwegian security experts, former diplomats, and NATO military planners reflected this sentiment in more than two dozen interviews conducted by the author between 2022 and 2023.

29 *"It's kind of dualistic":* Øyvind Nordsletten (former Norwegian ambassador to Russia), phone interview with author, August 2023.

FOUR DAYS AT THE ARCTIC CIRCUS

31 *eruption alerts have trickled:* "Volcanic Eruption Lights Up Iceland After Weeks of Earthquake Warnings—A Geologist Explains What's Happening," PreventionWeb, news release, December 19, 2023, https://www.preventionweb.net/news/volcanic-eruption-lights-iceland-after-weeks-earthquake-warnings-geologist-explains-whats.

31 *the famous Blue Lagoon is closed:* Catharine Fulton, "From Iceland: Magma Collecting Near Blue Lagoon, Svartsengi Power Plant," *Reykjavík Grapevine,* November 1, 2023, https://grapevine.is/news/2023/11/01/magma-collecting-near-blue-lagoon-svartsengi-power-plant.

32   *will soon be evacuated:* Jon Henley, "Residents of Volcano-Threatened Icelandic Town Allowed Brief Visit Home," *The Guardian,* November 12, 2023, https://www.theguardian.com/world/2023/nov/12/iceland-experts-predict-feared-volcanic-eruption-could-destroy-town-near-reykjavik.

32   *Silicon Valley of Cod:* Greg Harris, "Iceland's 'Silicon Valley of Cod' Holds Secrets for New England's Fishing Industry," *Boston Globe,* February 1, 2024, https://www.bostonglobe.com/2024/02/01/magazine/lessons-from-icelands-silicon-valley-of-cod.

32   *one swimming pool for every:* Egill Bjarnason, "Pool-landia," *Hakai,* April 11, 2017, https://hakaimagazine.com/features/pool-landia.

32   *death rates from drowning:* V. Rafnsson and H. Gunnarsdottir, "Fatal Accidents Among Icelandic Seamen: 1966–86," *Occupational & Environmental Medicine* 49, no. 10 (1992): 694–99, https://doi.org/10.1136/oem.49.10.694.

32   *the 1-10-1 myth:* "The 1-10-1 Myth," National Center for Cold Water Safety, information sheet, n.d., https://www.coldwatersafety.org/1-10-1-myth. Note: This has been disputed.

32   *one wilderness survival instructor:* While living in Reykjavík, I attended a ten-day Wilderness First Responder certification course offered by the National Outdoor Leadership School.

33   *its sidewalks heated by those:* Erik Pomrenke, "Does Reykjavík Have Heated Sidewalks?" *Iceland Review,* blog, March 15, 2023, https://www.icelandreview.com/ask-ir/does-reykjavik-have-heated-sidewalks.

33   *comes from renewable energy:* "Energy," Ministry of the Environment, Energy, and Climate Change, Government of Iceland, information sheet, n.d., https://www.government.is/topics/business-and-industry/energy.

34   *winds cancel dozens of flights:* Trine Jonassen, trans. Birgitte Annie Hansen, "The Arctic Convenes in a Restless World," *High North News,* October 23, 2023, https://www.highnorthnews.com/en/arctic-convenes-restless-world.

34   *has sent its largest presence in years:* This information comes from internal memos, U.S. State Department, obtained through a Freedom of Information Act request.

34   *scandals over reeducation programs:* Antonio Voce, Leyland Cecco, and Chris Michael, "'Cultural Genocide': The Shameful History of Canada's Residential Schools—Mapped," *The Guardian,* September 6, 2021, https://www.theguard ian.com/world/ng-interactive/2021/sep/06/canada-residential-schools-indig enous-children-cultural-genocide-map.

34   *does not live in remote arctic territory:* Some 75 percent of Canada's population lives within 100 miles of the nation's border with the United States. See "Canada: Quick Facts," Arctic Council, information sheet, n.d., https:// arctic-council.org/about/states/canada.

35   *are leery that those aims:* Heljar Havnes Seland and Johan Martin, "The Increasing Security Focus in China's Arctic Policy," Arctic Institute, Center for Circumpolar Security Studies, news release, July 16, 2019, https://www .thearcticinstitute.org/increasing-security-focus-china-arctic-policy.

35   *nearly twenty years:* The forces officially departed in 2006, but now make frequent stops to refuel nuclear submarines, or base spy planes, like the *Poseidon,* there.

35   *has its own ice-class specifications:* American Bureau of Shipping, *ABS Notations and Symbols,* March 2025, https://ww2.eagle.org/content/dam/eagle /rules-and-resources/RuleManager2/class-notations-table.pdf.

36   *Polar Code classes 1 through 7:* A great visualization of this comes from the U.S. Coast Guard, which distributes an infographic listing the "Homeports of Major Polar Icebreakers." Go to https://kennethrrosen.com/arctic.

36   *this designation is rather new:* "History Files (HF) and Technical Background (TB) Documents for Unified Requirements (URs)," International Association of Classification Societies Ltd., n.d., https://web.archive.org/web/20151019214521 /http://www.iacs.org.uk/document/public/Publications/TBs/URTB.pdf.

36   *the bulk of other international icebreakers:* Few of these vessels are military in nature. Some fall under a nation's coast guard, which may or may not be a division of that nation's national defense or military, while others are owned by private universities or research institutions or may even be cruise ships. Most countries have planned to build more ice-capable vessels, but historically those efforts have either stalled (as is the case in the U.S. and Russia)

or have been scuttled altogether (as in Canada). Most icebreakers operating today were built between 1954 (Finland's *Voima*, launched that year and retrofitted in 1979) and 2002 (KV *Svalbard*). The Kystvakt, like its American counterpart, had ordered more ice-capable ships.

36   *whose representative recapitulates:* Astri Edvardsen, trans. Birgitte Annie Hansen, "The United Arab Emirates at the Arctic Circle: 'Past Climate Promises Must Be Upheld,'" *High North News,* October 20, 2023, https://www.high northnews.com/en/united-arab-emirates-arctic-circle-past-climate-promises -must-be-upheld.

37   *as to catch a glimpse of:* Multiple attempts were made to speak with Senator Murkowski in person, by phone, and email, between October 2023 and December 2024.

37   *was not lost on the Japanese:* I came across Otis Hays Jr., *Alaska's Hidden Wars: Secret Campaigns on the North Pacific Rim* (University of Alaska Press, 2004) while traveling south of Coldfoot, Alaska. The slim edition by Hays, an intelligence officer in Alaska during the Second World War, offers a stunning history of how the national interests of three nations converged in the North Pacific, underscoring the region's enduring importance.

37   *tested its first hydrogen bomb:* "Tsar Bomba: The Largest Atomic Test in World History," National WWII Museum, New Orleans, August 29, 2020, https:// www.nationalww2museum.org/war/articles/tsar-bomba-largest-atomic -test-world-history.

37   *the creation of an early-warning system:* Gary Weir, "The DEW Line–Cold War Defense at the Top of the World," National Geospatial-Intelligence Agency, Medium, https://medium.com/@NGA_GEOINT/the-dew-line-cold-war-de fense-at-the-top-of-the-world-fbafdd90a542.

37   *followed by the founding of:* "NORAD Agreement," North American Aerospace Defense Command, fact sheet, n.d., https://www.norad.mil/About -NORAD/NORAD-Agreement.

37   *glided beneath the geographic:* "Nautilus (SSN-571)," Naval History and Heritage Command, n.d., https://www.history.navy.mil/browse-by-topic/ships /submarines/uss-nautilus.html.

NOTES

37   *the Third Reich once maintained its Atlantic Wall:* Earl F. Ziemke, *The German Northern Theater of Operations, 1940–1945* (Department of the Army, 1959); Allen F. Chew, *Fighting the Russians in Winter: Three Case Studies* (Combat Studies Institute, U.S. Army Command and General Staff College, 1981); Jack Adams, *The Doomed Expedition: The Norwegian Campaign of 1940* (Leo Cooper, 1989).

38   *for the circumpolar region:* "High north, low tension," is a Norwegian slogan meant to promote a form of soft cooperation among the eight arctic states.

38   *10,000-foot ice core:* For further reading on this, its scientific importance, and America's forays into Greenland, check out Paul Bierman, *When the Ice Is Gone: What a Greenland Ice Core Reveals About Earth's Tumultuous History and Perilous Future* (W. W. Norton, 2024).

39   *permanent fund model:* Formed nearly two decades later, Norway's sovereign wealth fund has ballooned well past Alaska's own, despite its relative youth. Whereas Alaska focused on the immediate fiscal benefits of its oil and gas industry, Norway saved toward a post-oil future.

39   *"arctic exceptionalism":* In Senator Murkowski's own words, the arctic was once and could still remain an "exceptional" place of cooperation and global benefit, "if only we try." Lisa Murkowski, "Arctic Exceptionalism," *Foreign Service Journal,* May 2021, https://afsa.org/arctic-exceptionalism.

39   *Russia began to modernize:* "Dark Arctic: NATO Allies Wake Up to Russian Supremacy in the Region," Reuters, November 16, 2022, https://www.reuters.com/graphics/ARCTIC-SECURITY/zgvobmblrpd/.

39   *testing of hypersonic munitions:* Jyri Lavikainen, "Strengthening Russia's Nuclear Forces in the Arctic: The Case of the Kinzhal Missile," Center for Strategic & International Studies, September 14, 2021, https://www.csis.org/analysis/strengthening-russias-nuclear-forces-arctic-case-kinzhal-missile.

39   *China had sought funding:* John Acher and Mette Fraende, "Greenland's Minerals Loom in China-Denmark Ties," Reuters, June 16, 2012, https://www.reuters.com/article/world/greenlands-minerals-loom-in-china-denmark-ties-idUSBRE85F0FC.

39   *Turkey began the process to join the treaty:* "Türkiye Set to Raise North Pole Stake with Key Spitsbergen Treaty," *Daily Sabah,* October 9, 2022, https://www.dailysabah.com/turkey/turkiye-set-to-raise-north-pole-stake-with-key-spitsbergen-treaty/news.

40   *Polar Silk Road:* Jane Nakano and William Li, "China Launches the Polar Silk Road," Center for Strategic & International Studies, February 2, 2018, https://www.csis.org/analysis/china-launches-polar-silk-road.

40   *spending across Scandinavia:* Many NATO members were encouraged to meet their 2 percent of national GDP spending on defense since the first Trump administration, but the numbers seen after the war in Ukraine began easily have surpassed those goals. Thomas Nilsen, "Norway to Hike Defense Spendings in Mid-Year Revised Budget," *Barents Observer,* May 2, 2024, https://www.thebarentsobserver.com/security/norway-to-hike-defense-spendings-in-midyear-revised-budget/165371; Thomas Nilsen, "With Northern Focus, Finland Applies EU Funding to Improve Military Mobility," *Barents Observer,* September 13, 2023, https://thebarentsobserver.com/en/security/2023/09/northern-focus-finland-applies-big-eu-funding-improve-military-mobility.

40   *historic fluctuations in volunteer applications:* Elisabeth Braw, "Russia Turns Swedish Home Guard into a Recruitment Triumph," *Politico,* February 6, 2024, https://www.politico.eu/article/russia-turns-swedens-home-guard-turns-recruitment-success.

40   *forced to reevaluate:* Two of the Royal Danish Navy patrol vessels dedicated to ensuring Greenland's security were lying idle in the Nuuk harbor for much of 2023–24. The government admitted it had neglected the nation's defense for years. Jacob Gronholt-Pedersen, Louise Rasmussen, Stine Jacobsen, and Jacob Gronholt-Pedersen, "Denmark Says It Has Neglected Greenland Defence for Years," Reuters, January 9, 2025, https://www.reuters.com/world/us-says-it-has-no-plans-increase-military-presence-greenland-2025-01-09. In January 2025, the government in Copenhagen announced a $1.5 billion defense package to include more sled-dog patrols. Jacob Gronholt-Pedersen and Jacob Stine, "Denmark Plans New Ships, Dog Sled Patrols in Greenland as Trump Seeks Con-

trol," Reuters, January 10, 2025, https://www.reuters.com/world/denmark-plans -new-ships-dog-sled-patrols-greenland-trump-seeks-control-2025-01-10.

40   *a representative from:* Author interview with unnamed representative, Canada Department of National Defence, December 2023, Nuuk, Greenland.

41   *installations across the arctic and subarctic:* Department of Defense Office of the Inspector General, *Evaluation of the Department of Defense's Efforts to Address the Climate Resilience of U.S. Military Installations in the Arctic and Sub-Arctic,* April 13, 2022, https://www.arctic.gov/uploads/assets/DODIG -2022-083.pdf.

41   *across governmental departments:* The U.S. Department of Defense released a revised arctic strategy implementation plan in July 2024.

41   *one former diplomat:* Author interview with unnamed diplomat, summer 2023, Vilnius, Lithuania.

42   *Founded in 2013:* "Chairman of Arctic Circle," Arctic Circle, fact sheet, n.d., https://www.arcticcircle.org/chairman-of-arctic-circle.

42   *attends Arctic Council meetings only virtually:* I am told that the Arctic Council states still communicate via written handouts, but that work is slow and burdensome. Jennifer Spence and Hannah Chenok, *The Future of Arctic Council Innovation: Charting a Course for Working-Level Cooperation,* Harvard University, Belfer Center for Science and International Affairs, report, February 20, 2024, https://www.belfercenter.org/publication/future-arctic -council-innovation-charting-course-working-level-cooperation.

43   *At that session:* Sara Olsvig (chair of the Inuit Circumpolar Council), follow-up interview with the author, December 2023, Nuuk, Greenland.

43   *as "lúpína" in Icelandic—was meant:* A great history of Iceland I found wildly enjoyable is Egill Bjarnason, *How Iceland Changed the World: The Big History of a Small Island* (Penguin, 2021).

"COMMUNISM IS OUR GOAL"

45   *many modern-day gulag-style:* Aleksei Navalny, the Russian rights activist and opposition leader, was found dead in one such camp in 2024. Valerie

Hopkins and Andrew Kramer, "Aleksei Navalny, Russian Opposition Leader, Dies in Prison at 47," *New York Times,* February 16, 2024, https://www .nytimes.com/2024/02/16/world/europe/aleksei-navalny-dead.html; Ivan Nechepurenko and Anton Troianovski, "Aleksei Navalny Found in Remote Arctic Prison, Easing Fears over His Safety," *New York Times,* December 25, 2023, https://www.nytimes.com/2023/12/25/world/europe/russia-navalny -found-prison.html.

45   *frontline cannon fodder:* "Russia's Ethnic Minorities Disproportionately Die in the War in Ukraine," *PBS News Hour,* December 11, 2023, https://www .pbs.org/newshour/show/russias-ethnic-minorities-disproportionately -conscripted-to-fight-the-war-in-ukraine.

46   *homestead plots and cash:* Neil MacFarquhar and Milana Mazaeva, "Russia Showers Cash on Men Enlisting in Ukraine War, Bringing Prosperity to Some Towns," *New York Times,* November 2, 2024, https://www.nytimes .com/2024/11/02/world/europe/russia-ukraine-war-draft.html.

47   *drone attacks reaching Murmansk:* Thomas Nilsen, "War in the Arctic," *Barents Observer,* September 16, 2024, https://thebarentsobserver.com/en/opin ions/2024/09/war-arctic. Russia has also launched several drone strikes against Kyiv that originated from the arctic city. Atle Staalesen, "Murmansk Sends Deadly New Year Greeting to Kyiv," *Barents Observer,* January 2, 2024, https:// www.thebarentsobserver.com/security/murmansk-sends-deadly-new-year -greeting-to-kyiv/163885.

47   *worked with its partners:* Maria Lagutina and Yana Leksyutina, "BRICS Countries' Strategies in the Arctic and the Prospects for Consolidated BRICS Agenda in the Arctic," *Polar Journal* 9, no. 1 (2019): 45–63, https://doi.org/1 0.1080/2154896X.2019.1618559.

47   *adjusting their behavior based on:* Malte Humpert, "Ukraine Petitions Norway, US, and EU for New Sanctions Against Russian Gas," *High North News,* February 6, 2024, https://www.highnorthnews.com/en/ukraine-petitions-nor way-us-and-eu-new-sanctions-against-russian-gas.

47   *is not international cooperation:* Elizabeth Buchanan's *Red Arctic: Russian Strategy Under Putin* (Brookings Institution Press, 2023) is an extremely misleading text

of Russian propaganda, insightful only for its attempt to frame Russian thinking about its arctic ambitions and ceaseless operations wreaking global instability.

47   *since Josef Stalin's "Red Arctic" propaganda:* Much of Russian President Vladimir Putin's approach to the arctic is modeled on the same ethos as his predecessor. Heather A. Conley, Matthew Melino, and Jon B. Alterman, "The Ice Curtain: Russia's Arctic Military Presence," Center for Strategic & International Studies, March 26, 2020, https://www.csis.org/analysis/ice-curtain-russias -arctic-military-presence.

48   *Artur Chilingarov, a former deputy chairman:* Scott G. Borgerson, "Arctic Meltdown: The Economic and Security Implications of Global Warming," *Foreign Affairs,* March 2, 2008, https://www.foreignaffairs.com/articles/arc tic-antarctic/2008-03-02/arctic-meltdown. Chilingarov passed away in June 2024, "Artur Chilingarov, Russian Polar Scientist and Member of Parliament, Dies at 84," Reuters, June 1, 2024, https://www.reuters.com/world/europe /artur-chilingarov-russian-polar-scientist-member-parliament-dies-84 -2024-06-01.

48   *trash-strewn and barely inhabited:* Big Diomede and Little Diomede straddle the international date line and are just 2.33 miles apart—close enough that, when the Bering Sea freezes, the villagers can walk between the two countries with ease.

48   *naval and air force maneuvers:* "U.S. Coast Guard Encounters People's Republic of China Military Naval Presence in Bering Sea," United States Coast Guard News, press release, July 10, 2024, https://www.news.uscg.mil/Press-Releases /Article/3834722/us-coast-guard-encounters-peoples-republic-of-china-mili tary-naval-presence-in.

48   *Russia's territory in North America:* A splendid modern history tied to historical inflection points is Charles Emmerson's *The Future History of the Arctic* (PublicAffairs, 2020).

49   *which critics decried:* The purchase was known then as "Seward's Folly." "U.S. Senate: Charles Sumner and the Purchase of Alaska," United States Senate, information sheet, n.d., https://www.senate.gov/about/powers-pro cedures/treaties/sumners-alaskan-project.htm.

49   *"our Hyperborean Eldorado":* "Speech of the Honorable Charles Sumner, of Massachusetts, on the Cession of Russian America to the United States," U.S. Senate, April 9, 1867, https://hdl.handle.net/2027/uiuo.ark:/13960 /t9086cf0p.

49   *seventy years after the purchase:* Indiana Democratic Senator Joe Donnelly's farewell address on the Senate floor, December 11, 2018, https://www.c-span .org/clip/us-senate/senator-joe-donnelly-delivers-farewell-address/4766785.

50   *that acreage may grow:* Farming in Alaska is a love-hate affair. In the late seventies, Alaska attempted to kickstart agriculture outside Fairbanks, inspiring a northward migration. But a broader "Farm Plan" did not bear out. Timothy Egan, "Alaska Farm Plan Yields Bitter Harvest," *New York Times,* March 4, 1992, https://www.nytimes.com/1992/03/04/us/alaska-farm-plan -yields-bitter-harvest.html. With warming temperatures, the state may do well to revisit its arable land potential. Yasmin Tayag, "The Surreal Abundance of Alaska's Permafrost Farms," *The New Yorker,* August 30, 2022, https://www .newyorker.com/news/annals-of-a-warming-planet/the-surreal-abundance -of-alaskas-permafrost-farms.

50   *literally falling into:* Rick Bowmer and Mark Thiessen, "Climate Change Destroyed an Alaska Village. Its Residents Are Starting Over in a New Town," AP News, September 26, 2024, https://apnews.com/article/climate-change-permafrost-melt ing-alaska-newtok-relocation-moving-292694f057b75f75a9438c794853ee25.

51   *while three-quarters of the:* Janis Kluge and Michael Paul, "Russia's Arctic Strategy Through 2035," Stiftung Wissenschaft und Politik, report, n.d., https:// www.swp-berlin.org/publikation/russias-arctic-strategy-through-2035.

51   *Chukotka region has shrunk by:* A. A. Dudarev, V. S. Chupakhin, and J. Ø. Odland, "Health and Society in Chukotka: An Overview," *International Journal of Circumpolar Health* 72 (2013): 20469, https://doi.org/10.3402/ijch.v72i0.20469.

51   *Canadian firm:* "Canada's Kincross Gold Sells Russia Assets at Half Price," Reuters, June 15, 2022, https://reuters.com/article/business/canadas-kinross -gold-sells-russia-assets-at-half-price-idUSKBN2NW14E.

51   *by an Australian firm:* Camille Lin, "The Companies Are Destroying the Environment and Harming the Local People in Chukota," *Polar Journal,*

October 29, 2024, https://www.polarjournal.net/these-companies-are-de
stroying-the-environment-and-harming-the-local-people-in-chukotka.

51   *seek to roll back environmental legislation:* Atle Staalesen, "Russian Oil Companies Seek to Soften Environmental Law Ahead of a Big Push into the Arctic,"
Arctic Today, February 2, 2021, https://www.arctictoday.com/russian-oil-compa
nies-see-to-soften-environmental-law-ahead-of-a-big-push-into-the-arctic.

51   *Moscow embraces these requests:* Atle Staalesen, "Russia Might Lower Environmental Standards in Arctic," *Barents Observer,* April 6, 2023, https://
www.thebarentsobserver.com/climate-crisis/russia-might-lower-environ
mental-standards-in-arctic/108440.

51   *marine mammals trained as spies:* Hvaldimir was hungry for life. Outgoing
and charismatic, Hvaldimir, a young beluga whale, also thirsted for attention.
He liked people. He roamed freely across arctic waters. First discovered off
the Norwegian coast on April 26, 2019, the beluga whale was 14 feet long and
roughly 2,700 pounds, a bit shy of the average for a normal adult male whale. A
poll conducted by the Norwegian Broadcasting Corporation received 25,000
respondents, which created a portmanteau from the Norwegian word for
whale and the name of the Russian president. He was named by consensus.

The public and researchers had found the name fitting if only because of
what they discovered when the whale followed a research boat to the northern Norwegian town of Hammerfest, on a crescent of land along an inlet
toward the Arctic Ocean. The ocean was where many cetaceans typically
live and hunt, remaining far from the shorelines. Hvaldimir was wearing
what seemed to be a harness, its buckles stamped in Cyrillic, translating
to "Equipment St. Petersburg." He was also unusually calm around people.
He was personable. He played fetch. He returned cell phones that had been
dropped into the central harbor in Hammerfest. He displayed a particular
ability to wrap cords and rope around the propellers of boats.

Hvaldimir was a conscript of the Russian Navy. Having spent his life
in captivity, trained to perform tasks like tagging enemy vessels and divers
or sniffing out mines and underwater surveillance equipment, Hvaldimir
would die without human intervention. He would not be able to reintegrate

into aquatic society after being incarcerated for most of his adolescence. Researchers estimated him to be between twelve and twenty years old.

Hvaldimir's prison had been Polyarny, a naval hub shrouded in secrecy on Russia's Kola Peninsula, home to hexagonal pens encased in rusted metal bars. There, under the watchful eyes of trainers and military officers, he honed skills foreign to his wild counterparts. His missions, like those of his American brethren in the U.S. Navy's Marine Mammal Program and Russia's Northern Fleet, were tactical, not playful—built around detection and infiltration.

Inside the stark military pens, belugas like Hvaldimir were fitted with cameras and sonar devices. They became the unseen guardians of naval bases, patrolling the hidden depths for intruders and mines, their intelligence put to use in ways humans could not match. Like dolphins and seals drafted into similar programs around the globe, Hvaldimir had his destiny tied to the ambitions of nations. These animals, far from the open seas, were warriors in a secret war beneath the waves. They silently passed the solitary chips of ice floating southward, first as large azure hexagons, then as smaller trapezoids, seemingly alive, respiring, animated by the winds or the wake of the passing ship. Pressure ridges—those hulking sheets of blue-green ice three feet or more thick—collided overhead as they swam silently beneath them. Like the icebergs, Hvaldimir was drifting toward certain demise.

Russia's Northern Fleet in particular had long used marine mammals like Hvaldimir as part of a covert effort to bolster its naval supremacy in the arctic. Seals, dolphins, and whales—creatures renowned for their intelligence and agility—became assets in a growing network of underwater espionage. In some ways, it seemed like Russia thirsted for its own share of attention in the arctic. Now, as arctic competition intensified, these animals were being called to a new theater of operations, where cold waters and the deafening silence of sub-zero depths hid mines, enemy divers, and—sometimes—whales in harnesses.

Hvaldimir was not just a product of Russian ambition but a symbol of something larger, more insidious: how the Arctic, once a vast and silent wilderness, was now being swallowed by the invisible hands of power. He was part of an arsenal that patrolled beneath the ice, where whales were conscripts, where nations whispered and the waters betrayed no one. The struggle for control, for territory, for dominance, played out beneath the surface, unseen and unfelt, like the slow creep of an oncoming storm.

Maybe Hvaldimir was merely something innocuous, helpful, like a service animal, trained to provide therapy to children. If Hvaldimir had been a spy, he was not alone. Spies, always fascinating, were being discovered everywhere in the high latitudes. Chinese and Russian diplomats were being removed from their posts in foreign capitals. Years passed. Hvaldimir became something of a local celebrity, and with that stardom came the poking and prodding of curious individuals. Malice may not have been their intent, but dangerous things were placed into his mouth. He was often injured by boat propellers, his body gashed by sharp objects. Then, in September 2024, Hvaldimir's body was discovered, lifeless, floating in the waters of his adopted home. First, a researcher thought he had been shot. Animal-rights groups called for investigations. An autopsy under direction from the Norwegian Directorate of Fisheries revealed what had been thought to be a bullet was, in fact, "completely superficial."

Hvaldimir died of what seemed to be organ failure. He died of hunger, of relying on other hands to feed him. When he died—quietly, anonymously—it was as if the arctic itself had swallowed him whole. Not from gunfire or violence, but from the simplest betrayal of all: neglect. A spy without a mission, a soldier without a war. He had outlived his usefulness, no longer an asset, just another casualty in a cold, hidden conflict. He died not as a whale but as a ghost of human ambition in the arctic, another evanescent presence in a land filling with spies. Ferris Jabr, "The Whale Who Went AWOL," *New York Times Magazine,* January 14, 2024, https://www.nytimes.com/2024/01/14/magazine/hvaldimir-whale.html; Hanne Bernhardsen Nordvåg, "Folket har talt—hvalen skal hete Hvaldimir," NRK, May 3, 2019, https://www.nrk.no/tromsogfinnmark/folket-har-talt-_-hvalen-skal-hete-hvaldimir-1.14536969; A. N. Walnum, "Trodde telefonen var tapt. Så kom "«Hvaldimir» til unnsetning," *Dagbladet,* May 5, 2019, https://www

.dagbladet.no/nyheter/trodde-telefonen-var-tapt-sa-kom-hvaldimir-til-unn setning/71031243; Thomas Nilsen, "Satellite Images Reveal Russian Navy's Secret Arctic Marine Mammal Facility," *Barents Observer,* May 27, 2019, https://thebarentsobserver.com/en/security/2019/05/here-northern-fleets -secret-marine-mammal-program; Darko Janjevic, "Russian 'Spy Whale' May Have Been Therapist," *DW,* May 8, 2019, https://www.dw.com/en/rus sian-spy-whale-may-have-provided-therapy-for-children/a-48660795.

51 *was to return to Spitsbergen:* I arrived in Spitsbergen in October 2022 and returned twice.

52 *Sanctions prevent the barges from coming ashore:* Anastasia Tenisheva, "Russia Hits Out at Norway over Blocked Arctic Archipelago Access," *Moscow Times,* June 29, 2022, https://www.themoscowtimes.com/2022/06/29/russia -hits-out-at-norway-over-blocked-arctic-archipelago-access-a78138.

53 *have dozens of words:* This is generally true, and some are beautiful. See Lucien Schneider, *Ulirnaisigutiit: An Inuktitut-English Dictionary of Northern Quebec, Labrador and Eastern Arctic Dialects,* trans. Dermot Collis (Les Presses de L'Universitie Laval, 2022). For example, *qanik* for snow falling, *aniu* for snow used to make water, *pukak* for crystalline snow on the ground, *aputi* for snow on the ground, and *qinu* for slushy ice by the sea.

53 *During a meeting of the Communist Party:* Chris Buckley, Keith Bradsher, Vivian Wang, and Austin Ramzy, "China's Leader Strikes a Defiant Note, Warning of 'Stormy Seas,'" *New York Times,* October 16, 2022, https://www .nytimes.com/2022/10/16/world/asia/china-congress-xi-jinping.html.

54 *to revitalizing Barentsburg and Pyramiden:* Astri Edvardsen, "Russia Plans Grand Upgrade of Svalbard Infrastructure," *High North News,* September 29, 2022, https://www.highnorthnews.com/en/russia-plans-grand-upgrade-svalbard-in frastructure.

54 *along with its partners in BRICS:* The nations in BRICS are: Brazil, Russia, India, China, South Africa, Egypt, Ethiopia, Indonesia, Iran, and the United Arab Emirates.

54 *reiterates the security situation:* Hilde-Gunn Bye, "Norwegian Prime Minister: 'No Signs of an Increased Security Threat in the North,'" *High North News,*

October 18, 2022, https://www.highnorthnews.com/en/norwegian-prime-min ister-no-signs-increased-security-threat-north.

"WE HAVE TIME"

58   *Beginning in 1940:* The proprietor of Atuagkat Bookstore in Nuuk supplied me with a handful of books about the American occupation and presence in Greenland.

59   *the military could also store:* Karl and Bernhard Philberth were physicists and priests who considered Greenland's ice sheet as a possible storage site for nuclear waste in the 1950s. Their plan was to:

- Reprocess spent reactor fuel to recycle long-lived nuclides
- Fuse short-lived radionuclides into glass or ceramic
- Surround the waste with lead and transport it to Greenland
- Scatter the waste over a small area of the ice sheet, far from the coast

59   *the cabin used by Knud Rasmussen:* The cabin is now in Qaanaaq, and is home to incredible treasures.

60   *The base at Thule:* This is one of the stranger things done to make amends for past sins: renaming a place to honor those who were displaced, giving them nothing but a knowing nod. In Alaska, for example, the town of Barrow was renamed Utqiagvik. The town didn't even want that to happen. In Greenland, the United States was approached by the Kingdom of Denmark (both Denmark and Green- land) and asked if the United States would consider changing the base name during the transition from the U.S. Air Force to the U.S. Space Force. Pituffik (bee-doo-FEEK) is the traditional Greenlandic name for the region where the base is located, and after consultation among all three governments, that name was selected to pay homage to the base's ties to the Greenlandic people and cul- ture. As a result, the base was renamed to recognize Greenlandic cultural heritage.

60   *dogsled patrol that travels:* They are called the Sirius Patrol, about which there are many Danish shows, articles, and documentaries.

60   *is home to reindeer:* A friend, Tommy Sandal, whom I met on a distant spit of land in Spitsbergen, pointed me toward his grandma's blood sausage recipe:

- 6 liters of blood (four reindeer)
- milk (powdered)
- 1 kg of sugar
- 4 packs of cardamom
- 4 packs of allspice
- 4 packs of cinnamon
- vanilla (to taste)
- 5 kg wheat flour
- (maybe some raisins or reindeer fat in strips)

62   *makes the perimeter penetrable:* Documents, photos, and materials not present in the publicly available report were obtained through a Freedom of Information Act request. Copies of all documents are available at https://kennethrrosen.com/arctic.

62   *who took over command of:* There is a rotating cast of officials across the arctic, many of whom rely on aides or the local population for quick lessons before serving short-term tours.

63   *2022 National Strategy for the Arctic Region:* "Fact Sheet: The United States' National Strategy for the Arctic Region," The White House, October 7, 2022, https://bidenwhitehouse.archives.gov/briefing-room/statements-releases/2022/10/07/fact-sheet-the-united-states-national-strategy-for-the-arctic-region.

64   *wrote me in an email in 2022:* As per my email correspondence with an unnamed U.S. State Department representative, December 16, 2022.

64   *had already approved funding:* Mark Thiessen, "Cruising to Nome: The First U.S. Deep Water Port for the Arctic to Host Cruise Ships, Military," AP News, June 18, 2023, https://apnews.com/article/alaska-arctic-port-nome-china-russia-588201b311513709404344fbc0d0e913.

64   *would be seriously delayed:* Alex DeMarban, "$663M Arctic Port Delayed, Frustrating Nome Officials and Alaska Congressional Delegation," *Anchorage Daily News,* October 29, 2024, https://www.adn.com/business-economy/2024/10/29/663m-arctic-port-delayed-frustrating-nome-officials-and-alaska-congressional-delegation.

64   *reopening its own former base:* Dave Leval, "Sullivan: Navy Considering Reopening Base in Adak," *Alaska's News Source,* March 14, 2021, https://www .alaskasnewssource.com/2021/03/15/sullivan-navy-considering-reopening -base-in-adak.

65   *sought to open a deep-water:* "Road Network for Utqiaġvik, Atqasuk, and Wainwright Arctic Strategic Transportation and Resources Project North Slope, Alaska," Alaska Department of Natural Resources, April 2020, https:// dnr.alaska.gov/projects/astar/20_001_RoadNetworkfor_UTQ_ATQ_WAI _StudywithAppendixATechMemosFinal.pdf.

66   *has sought to gain a competitive advantage:* Rush Doshi, Alexis Dale-Huang, and Gaoqi Zhang, "Northern Expedition: China's Arctic Activities and Ambitions," Brookings Institution, April 2021, https://www.brookings.edu/wp -content/uploads/2021/04/FP_20210412_china_arctic.pdf.

67   *the IS18 satellite relay looms:* "IS18, Qaanaaq, Greenland, Denmark," Preparatory Commission for the Comprehensive Nuclear-Test-Ban Treaty Organization (CTBTO), Thumbnail Profile: Greenland, n.d., https://www.ctbto .org/our-work/station-profiles/is18-qaanaaq-greenland-denmark.

68   *"But look, it is no secret":* "Briefing on the Administration's Arctic Strategy," U.S. Department of State, April 23, 2020, https://2017-2021.state.gov /briefing-with-senior-state-department-official-on-the-administrations-arc tic-strategy.

68   *"for our enemies, we have":* This has been taken way out of context; even the Swedish Ambassador was ill-advised to use the quote. It is from a 1950s revolutionary song called "My Motherland" (朋友來了有好酒，若是那豺狼來了，迎接它的，有獵槍).

68   *pressured it to foot the bill:* "China Withdraws Bid for Greenland Airport Projects: Sermitsiaq Newspaper," Reuters, June 4, 2019, https://www.reu ters.com/article/business/china-withdraws-bid-for-greenland-airport-proj ects-sermitsiaq-newspaper-idUSKCN1T5190. In late 2024, the Danish government then revoked the airport's authorization for international flights, frustrating locals and officials in Greenland, who saw the projects as a means to bring more tourists and economic growth to their island nation.

68   *"a polar great power":* The expression is widely open to interpretation and can also refer to China's skills and capability.

69   *treasure direct foreign investment:* Rasmus Leander Nielsen and Maria Ackrén, *The Second Foreign and Security Policy Opinion Poll in Greenland,* December 2024, https://da.uni.gl/media/4bwb5ict/survey-report-leander_ ackre-n_final.pdf.

69   *prompting Nuuk to strike a mining deal:* Jacob Gronholt-Pedersen, "Greenland Gives Danish-French Group Permit to Mine Rock with Green Potential, in Wake of Trump Interest," Reuters, May 21, 2025, https://www.reuters .com/sustainability/climate-energy/greenland-gives-danish-french-group -permit-mine-rock-with-green-potential-wake-2025-05-21.

"A WARM WELCOME IN COLD WATERS"

71   *the 226-foot boat:* "HERMES (Fishing Vessel) IMO 9857535MMSI 9857535," n.d., https://www.marinetraffic.com/en/ais/details/ships/shipid:6387849/m msi:257640000/imo:9857535/vessel:HERMES.

71   *Three officers and myself board the:* The three officers were Lieutenants Johansen, Soløy, and Ocean, plus myself.

72   *vocabularies dedicated to the:* "Senja, a large island north of Lofoten, recorded thirty local words for various types of wind." From Arne Lie Christensen, *Det norske landskapet* (Pax, 2002), 75.

72   *on this day is near gale:* The Norwegians use the Beaufort Scale; the Americans use "sea state." The Beaufort Scale, named after Admiral Sir Francis Beaufort, lists rankings from 0 to 12, with 0 being calm waters and no wind speed or wave height and 12 representing hurricane-force winds and waves to match.

72   *do little to protect from overfishing:* Manuela Andreoni, "Counting All the Fish in the Sea May Be Even Trickier than Scientists Thought," *New York Times,* August 22, 2024, https://www.nytimes.com/2024/08/22/climate/fish -stocks-overcounting.html.

72   *for 44 percent of all unique:* Eye on the Arctic, "Fishing Vessels Dominate Arctic Shipping Traffic in 2022," *Barents Observer,* April 13, 2024, https://

www.thebarentsobserver.com/arctic/fishing-vessels-dominate-arctic-ship ping-traffic-in-2022/119032.

73   *expected to increase by sixfold:* U.S. Committee on the Marine Transportation System, CTR (CMTS), *A Ten-Year Projection of Maritime Activity in the U.S. Arctic Region, 2020–2030,* September 2019, https://oceanconservancy.org/wp-con tent/uploads/2021/08/CMTS_2019_Arctic_Vessel_Projection_Report.pdf.

73   *home to at least:* Brenda Norcross and Katrin Iken, "Hidden Ocean 2016: Chukchi Borderlands: Fishes in the Arctic," NOAA Ocean Exploration, n.d., https:// oceanexplorer.noaa.gov/explorations/16arctic/background/fishes/fishes.html.

73   *species of marine mammals:* "Marine Mammals," Arctic Ocean Biodiversity, Census of Marine Life, information page, n.d., http://www.arcodiv.org /MarineMammals.html.

73   *almost twenty years later:* Ying Zhang, Siwa Msangi, James Edmonds, and Stephanie Waldhoff, "Limited Increases in Arctic Offshore Oil and Gas Production with Climate Change and the Implications for Energy Markets," *Scientific Reports* 14, no. 1 (2024): 6699, https://doi.org/10.1038/s41598-024-54007-x.

73   *U.S. Geological Survey from 2008:* U.S. Geological Survey, *Circum-Arctic Resource Appraisal: Estimates of Undiscovered Oil and Gas North of the Arctic Circle,* July 24, 2008, https://www.usgs.gov/publications/circum-arctic -resource-appraisal-estimates-undiscovered-oil-and-gas-north-arctic.

73   *up to fifteen feet thick in some areas:* "Sea Ice," National Snow and Ice Data Center, information page, n.d., https://nsidc.org/learn/parts-cryosphere/sea-ice.

73   *ocean is slightly less than:* "Arctic Ocean," The World Factbook, June 10, 2025, https://www.cia.gov/the-world-factbook/oceans/arctic-ocean.

74   *It certainly will be by:* A. Jahn, M. M. Holland, and J. E. Kay, "Projections of an Ice-Free Arctic Ocean," *Nature Reviews: Earth & Environment* 5 (2024): 164–76, https://doi.org/10.1038/s43017-023-00515-9.

75   *start at $32,000:* Top of the World: North Pole Cruise, 56 Parallel, advertisement, n.d., https://www.56thparallel.com/russia-tours/north-pole-cruise.

75   *because there are few vessels:* Heiner Kubny, "Lack of Ships Delays Development of Northeast Passage," *Polar Journal,* September 21, 2023, https:// polarjournal.net/lack-of-ships-delays-development-of-northeast-passage.

75   *aboard the KV* Svalbard: My embarkation was from September into October 2023.

77   *U.S. Coast Guard had plans:* In December 2024, the coast guard purchased a commercial icebreaker capable of Arctic Ocean navigation. Its home port would be in Juneau, though the construction of the port would take "several years." For more information, see https://www.mycg.uscg.mil/News/Article /4016098/coast-guard-adds-first-polar-icebreaker-to-its-fleet-in-25-years.

77   *So the crew relies on:* "Earth Observation for Polar Monitoring: Foundation Models for Earth Observation," Polar View, n.d., https://polarview.org/news -press/foundation-models-for-earth-observation.

78   *radar system snoops on Russia's:* Andrew Higgins, "On a Tiny Norwegian Island, America Keeps an Eye on Russia," *New York Times,* June 13, 2017, https://www .nytimes.com/2017/06/13/world/europe/arctic-norway-russia-radar.html.

78   *will soon be stationed at:* Atle Staalesen, "Norway Establishes Arctic Base for Long-Range Drones," *Barents Observer,* April 3, 2024, https://www.thebar entsobserver.com/security/norway-establishes-arctic-base-for-longrange -drones/165666.

78   *An early-warning system against:* Atle Staalesen, "Norwegians, Americans Build Arctic Satellite Station Against Enemy Cruise Missiles," *Barents Observer,* April 11, 2024, https://www.thebarentsobserver.com/security/norwegians-amer icans-build-arctic-satellite-station-against-enemy-cruise-missiles/166394.

78   *nearly the exact location reached:* Parry's fifth expedition, in 1827, reached 82 degrees, 45 minutes north, setting a record for northern exploration that was not broken until 1875, by Albert Hastings. William Edward Parry, *Narrative of an Attempt to Reach the North Pole* (repr., Legare Street Press, 2023).

79   *Norway created a military buffer zone:* These self-imposed restrictions were under review as late as May 2025. Thomas Nilsen, "Norway Eases Self-Imposed Restrictions on NATO Training," *Barents Observer,* May 22, 2025, https:// www.thebarentsobserver.com/security/norway-eases-selfimposed-restric tions-on-nato-trainingnbsp/430295.

79   *There are the guns:* The cannon during my embarkation was a 40mm, but the vessel was originally equipped with a 57mm cannon.

80   *explosions like Howitzer salvos:* This line is a hat tip to Nansen, who in his book *Farthest North* describes the sounds of ice against the hull as "explosions like cannon salvoes." Fridjtof Nansen, *Farthest North: The Incredible Three-Year Voyage to the Frozen Latitudes of the North* (Modern Library, 2000).

81   *Though the ships and crews:* Four years before my embarkation aboard the two icebreakers, KV *Svalbard* steamed to the Bering Sea to retrieve instruments for the Coordinated Arctic Acoustic Thermometry Experiment, which had originally been planned for pickup by the *Healy*. Even then, Russian media had called the voyage an intrusion, noting that an "armed NATO warship" had invaded its waters.

81   *The National Science Foundation:* Cuts to the NSF's Polar Program have hamstrung the agency's efforts in the region, thanks to cost-cutting efforts by President Trump and Elon Musk's Department of Government Efficiency. Aatish Bhatia, Irineo Cabreros, Asmaa Elkeurti, and Ethan Singer, "Trump Has Cut Scientific Funding to Its Lowest in Decades," *New York Times,* May 22, 2025, https://www.nytimes.com/interactive/2025/05/22/upshot/nsf -grants-trump-cuts.html.

81   *The KV* Svalbard *retrieved the moorings:* It was in fact Lieutenant Soløy who dove below the ice to free one of the buoys.

81   *breaking four feet of ice:* The *Healy* can break ice up to 8 feet (2.44 meters) thick by using a ramming and reversing procedure.

82   *It is one of four:* Paul Berger and Daniel Michaels, "Texas Shipyard Deal Would Bring Arctic Icebreakers Trump Seeks," *Wall Street Journal,* June 11, 2025, https://www.wsj.com/articles/texas-shipyard-deal-would-bring-arctic -icebreakers-trump-seeks-7d9b6c8e. Increasing American industrial capacity in the realm of icebreakers is a win for the U.S., but the situation is dire. The authors incorrectly note that "the U.S. has just three Arctic-ready icebreakers in service." True, the coast guard recently purchased and reoutfitted an icebreaker, the U.S. Coast Guard Cutter *Storis*. But there are three others.

    The U.S. Coast Guard Cutter *Healy*, which services the Arctic Ocean proper and its surrounding waters, is constantly homeported and under serious repair. The U.S. Coast Guard Cutter *Polar Star* is the exclusive facilitator of American expeditions to Antarctica. And a third, the U.S. Coast Guard Cutter *Polar Sea*, is only used when cannibalizing it for parts to use aboard the *Polar Star*.

The new icebreaker, the *Storis*, will be homeported in Juneau, the first American icebreaker to be based in Alaska in decades. (The *Healy* is based in Seattle.) But the project to facilitate its base in Alaska's capital will take, according to the coast guard, "several years." Perhaps when the deep-sea port in Nome opens, if ever, the *Storis* can make its way closer to the Arctic Circle. Until then, it will be located far from some of the nation's busiest shipping lanes as warming temperatures in the north increase traffic and commerce through the Bering Strait.

The Canadians have a wonderful track record when it comes to building ice-faring ships, as does Finland. With this recent news and that of the U.S. signing its new ICE Pact with those two nations, perhaps, however sclerotic, Washington is finally taking the steps necessary toward reestablishing an impressive presence in the world's fastest-warming region. However, with a tenuous relationship between Trump and the Canadian government, it is not assured that the collaboration on icebreakers will remain.

82   *as part of the Nansen and:* This is according to an official press release from the U.S. Coast Guard, "U.S. Coast Guard Cutter *Healy*, Scientists Deploy Ice Stations," August 14, 2023, https://www.news.uscg.mil/Press-Releases/article/3491123/us-coast-guard-cutter-healy-scientists-deploy-ice-stations.

83   *in a thin layer of ice:* I would spend several nights aboard the *Healy* as it made its way to the Norwegian mainland.

84   *"operations in more than":* From an email to the Arctic Icebreaker Coordination Committee, dated September 3, 2024, with the subject line "Arctic Research Opportunity Onboard USCGC Healy, October 1st through December 15, 2024; Proposals due by September 10th."

85   *undertaken during the Cold War:* For further comparison between today's arctic and the arctic of the Cold War period, see Richard Petrow, *Across the Top of Russia: The Cruise of the USCGC Northwind into the Polar Seas North of Siberia* (David McKay, 1967).

85   *the* Healy *traversed the far:* A map of the route and the moorings can be found at https://www.highnorthnews.com/sites/default/files/styles/media _image/public/2023-10/Healys%20rute%20under%20NABOS-toktet%20 2023_Kodiak-Troms%C3%B8.jpg.

85   *Russia issues a notice-to-airmen:* Thomas Nilsen, "USCG *Healy* Docks in Tromsø after Joint Voyage with Norwegian Coast Guard Northeast of Svalbard," *Barents Observer,* October 1, 2023, https://www.thebarentsobserver .com/arctic/uscg-healy-docks-in-tromso-after-joint-voyage-with-norwegian -coast-guard-northeast-of-svalbard/143285.

86   *The chefs prepare a dinner of:* The official coast guard press release about the end of the mission reads, "The two ships transited together toward Tromsø while crew members participated in an exchange on each other's vessel to foster a deeper understanding of the other service's operation." See "U.S. Coast Guard Cutter *Healy* Departs Tromsø, Norway Following Engagements with Norwegian Coast Guard, Maritime Government Officials," U.S. Coast Guard News, October 6, 2023, https://www.news.uscg.mil/Press-Releases /Article/3549076/us-coast-guard-cutter-healy-departs-troms-norway-fol lowing-engagements-with-nor.

86   *vessel that has been tracking alongside:* Meaning "lying-to," in naval jargon.

87   *an ensign named Patrick:* I made a considerable effort while aboard the *Healy* to have an interview with the commanding and executive officers but was rebuffed until a phone call was arranged several months later.

88   *one coast guard official tells me:* Author's email exchange with an unnamed source, late 2024.

88   *skirmish between the Soviet Union and:* The *Northwind's* sister ship, the *Eastwind,* discovered a second attempt by Germans to set up a weather station in Greenland during World War II. There were many attempts by the Germans to reach Greenland. For further reading, see, among others, Stetson Conn, Rose C. Engelman, and Byron Fairchild, *Guarding the United States and Its Outposts* (Government Printing Office, 1964), 1–29; "A History of ATC in Greenland"; Ole Guldager's *Americans in Greenland in World War Two* and *Arctic Sun* from the *Greenland History* series; and Bernt Balchen,

Corey Ford, and Oliver La Farge, *War Below Zero: The Battle for Greenland* (Houghton Mifflin, 1944).

SCIENTISTS, RESEARCHERS, DIPLOMATS, SPIES

91   *along the Paatsjoki River:* Given the distance the river travels, it has several names in several dialects and languages. I've chosen here to use the Norwegian name.

91   *threat to the native Atlantic salmon:* Elizaveta Vereykina, "'We Have Declared a War on Pink Salmon': Norway Fights the Invasive Species in the Arctic," *Barents Observer,* June 30, 2023, https://thebarentsobserver.com/en/2023/06/we-have-declared-war-pink-salmon-norway-fights-invasive-species-arctic.

92   *blanches the wilderness around town:* Maria Antonova, "Balancing Growth and Environment," *Moscow Times,* July 25, 2008, https://web.archive.org/web/20080803173348/http://www.themoscowtimes.com/article/600/42/369177.htm.

93   kuksa *(hand-carved wooden cup):* The traditional *kuksa* was used by the Sámi for berry picking, drinking, and eating. It is a symbol of resilience and resourcefulness and can be seen being used by those traveling across northern Scandinavia and Greenland.

94   *there were authentic spies:* Anton Troianovski, "An Arctic Spy Mystery: An Arrest in Moscow Shakes Norway's Far North," *Washington Post,* February 3, 2018, https://www.washingtonpost.com/news/world/wp/2018/02/03/feature/an-arrest-in-moscow-leads-to-a-norwegian-espionage-mystery.

94   *congressional staff members were quietly discussing:* Information here comes from a Freedom of Information Act request to the State Department F-2023-09570, submitted on June 8, 2023. It revealed a trove of internal communications and highlighted the department's objectives in opening the mission. To review all files, go to https://kennethrrosen.com/arctic.

94   *was posted to the lone office:* The unnamed "watcher" and I met and spoke off the record for more than an hour about the new position, its objectives, and the work being done by the office.

95   *considered NATO's Achilles' heel:* J. Wither, "Svalbard: NATO's Arctic 'Achilles' Heel,'" *RUSI Journal* 163, no. 5 (2018): 28–37, 47.

95   *GPS signals were spoofed:* Thomas Nilsen, "Someone Is Messing with GPS Signals in Svalbard Airspace," *Barents Observer*, July 19, 2025, https://www.thebarentsobserver.com/news/someone-is-messing-with-gps-signals-in-svalbard-airspace/433658.

95   *arrived as a volunteer researcher:* Yasmin Sollerman, Ola Haram, and Oliver Bellinder, "PST om siktelsen: Har spionsiktedes russiske identitet," *VG*, October 28, 2022 (in Russian), https://www.vg.no/i/763KXw.

96   *Russian diplomats based in Norway and Sweden:* Officials and journalists were expelled at a rapid clip in the months after Russia invaded Ukraine. A selection of the coverage of those incidents follows: Jan Olsen, "Norway Expels 15 Russian Diplomats Suspected of Espionage," *PBS News,* April 13, 2023, https://www.pbs.org/newshour/world/norway-expels-15-russian-diplomats-suspected-of-espionage; "Russia Says Iceland 'Destroys' Ties by Suspending Embassy Operations," Reuters, June 10, 2023, https://www.reuters.com/world/europe/russia-says-iceland-destroys-ties-by-suspending-embassy-operations-2023-06-10; and "Sweden Expels a Chinese Journalist, Calling Her a Threat to National Security, Report Says," AP News, April 8, 2024, https://apnews.com/article/sweden-china-journalist-expelled-national-security-71c1339984ba197c10ab34af4877e098.

96   *One academic lost his post:* Lassi Heininen, "Background Information on the Case of the Termination of the Emeritus Professor Contract of Lassi Heininen," *Arctic Politics*, May 20, 2024, https://arcticpolitics.com/news/background-information-case-termination-emeritus-professor-contract-lassi-heininen.

96   *increasing hybrid warfare attacks:* Thomas Nilsen, "Intensified GPS Jamming Is Side Effect of Russia's Self-Protection of Kola Bases," *Barents Observer,* May 31, 2024, https://thebarentsobserver.com/en/security/2024/05/intensified-gps-jamming-side-effect-russias-self-protection-military-bases-kola.

96   *extending well beyond Scandinavia:* Selam Gebrekidan, "Electronic Warfare Confounds Civilian Pilots, Far from Any Battlefield," *New York Times,* November 21, 2023, https://www.nytimes.com/2023/11/21/world/europe/ukraine-israel-gps-jamming-spoofing.html.

97   *had long noted that:* CIA documents declassified in 1990 reveal the importance of the Norwegian border with the U.S.S.R. to national security. See https://www.cia.gov/readingroom/docs/CIA-RDP79T01018A000100040 001-6.pdf.

97   *inaugural Arctic Ambassador-at-Large:* Ambassador Sfraga was a regular at events focusing on the region and its security.

97   *emphasized that the position:* "Risch on SFRC Passage of Unqualified State Department Nominees," U.S. Senate, Foreign Relations Committee, press release, March 20, 2024, https://www.foreign.senate.gov/press/rep/release /risch-on-sfrc-passage-of-unqualified-state-department-nominees.

98   *"Greenland must be regarded as covered by":* C. L. Sulzberger, "The Greatest Empire Left," *New York Times,* August 16, 1975, https://www.nytimes .com/1975/08/16/archives/the-greatest-empire-left-foreign-affairs.html.

98   *one senior State Department official:* Author had several phone conversations with this unnamed State Department official between December 2022 and July 2024.

99   *the "Seven C's":* M. Sfraga and J. Durkee, eds., *Navigating the Arctic's 7Cs* (Polar Institute, 2021), https://www.wilsoncenter.org/sites/default/files/media /uploads/documents/Polar_7Csbook.pdf.

100   *had reservations about Dr. Sfraga:* Transcript of discussion regarding the nomination of Michael Sfraga, Executive Session, September 24, 2024, *Congressional Record* 170, no. 149 (2024): S6345–47, https://www.congress .gov/118/crec/2024/09/24/170/149/CREC-2024-09-24-pt1-PgS6345-2.pdf.

100   *each time only after being confronted:* Several attempts to reach Dr. Sfraga after his appointment did not materialize in a meeting or call.

101   *"He has been very clear":* Liz Ruskin, "For the First Time, America Has an Arctic Ambassador, and He's Alaskan," *KTOO,* September 24, 2024, https:// www.ktoo.org/2024/09/24/for-the-first-time-america-has-an-arctic-ambas sador-and-hes-alaskan.

102   *hearing in April 2025:* Anna Lionas, "Bering Strait Highlighted in Arctic Security Presentation," *The Nome Nugget,* April 10, 2025, https://www.north ernjournal.com/r/3784e857.

NOTES

THE LAW OF THE REALM

103   *before boarding a Super King:* Nicholai interview.

104   *the airline covers the housing:* from interviews with airport staff at Constable Point.

105   *annual $511 million block:* Diana Roy and Jonathan Masters, "What Would Greenland's Independence Mean for U.S. Interests?," Council on Foreign Relations, March 13, 2025, https://www.cfr.org/article/greenlands-independence-what-would-mean-us-interests.

105   *Chinese firm proposed:* "China Withdraws Bid for Greenland Airport Projects: Sermitsiaq Newspaper," Reuters, June 4, 2019, https://www.reuters.com/article/business/china-withdraws-bid-for-greenland-airport-projects-sermitsiaq-newspaper-idUSKCN1T5190.

105   *"Greenland is a black hole":* Karl-Peter interview, Constable Point, Greenland, December 2023.

105   *roughly $11,000 per citizen:* Simon Johnson, "Greenland's Future Must Be Decided by Greenland's People," *Project Syndicate,* March 3, 2025, https://www.project-syndicate.org/commentary/resistance-to-trump-effort-for-us-control-of-greenland-by-simon-johnson-2025-03.

106   *established in the 1920s:* Tanguy Sandre, Jean-Paul Vanderlinden, Arjan Wardekker, and Jeanne Gherardi, "Living with 'Slow Upheavals': Unsettling a Resilience-Based Approach in Ittoqqortoormiit (Kalaallit Nunaat)," *Frontiers in Climate* 7 (2025), https://www.frontiersin.org/journals/climate/articles/10.3389/fclim.2025.1563320/full.

106   *no narwhal to hunt:* Sofia Moutinho and Regin Winther Poulsen, "The Last Hunt? Future in Peril for 'The Unicorn of the Sea,'" *The Guardian,* June 2, 2022, https://www.theguardian.com/environment/2022/jun/02/the-last-hunt-greenland-split-over-protection-for-the-narwhal.

106   *"this is our gold":* Nunduu and Karline interview.

107   *"we need Greenland for national":* Katherine Long and Alexander Ward, "U.S. Orders Intelligence Agencies to Step Up Spying on Greenland," *Wall Street Journal,* May 6, 2025, https://www.wsj.com/world/greenland-spying-us-intelligence-809c4ef2.

108   *transferring control of Greenland:* Brent Hardt, "Advancing U.S. Interests in Greenland," German Marshall Fund, January 16, 2025, https://www.gmfus .org/news/advancing-us-interests-greenland.

108   *"play an oversized role":* Danish diplomat interview, Copenhagen, Denmark, 2023.

108   *"U.S. interest phase China out":* Trine Jonassen, "Greenland Strips Chinese Mining Firm of License for Iron Ore Deposit," *High North News,* November 23, 2021, https://www.highnorthnews.com/en/greenland-strips-chinese-mining -firm-license-iron-ore-deposit.

108   *Greenland is not for sale:* "Denmark PM Repeats That Greenland Is Not for Sale," Reuters, February 23, 2025, https://www.reuters.com/world/denmark -pm-repeats-that-greenland-is-not-sale-2025-02-03.

108   *to replace Denmark's annual subsidy:* Maggie Haberman and Michael Crowley, "Inside Trump's Plan to 'Get' Greenland: Persuasion, Not Invasion," *New York Times,* April 10, 2025, https://www.nytimes.com/2025/04/10/us/poli tics/trump-greenland-denmark.html.

109   *so roundly shunned:* Kirsten Grieshaber, "Vance and Wife to Tour U.S. Military Base in Greenland After Diplomatic Spat over Uninvited Visit," *PBS News,* March 28, 2025, https://www.pbs.org/newshour/world/vance-and-wife-to-tour -u-s-military-base-in-greenland-after-diplomatic-spat-over-uninvited-visit.

109   *the math of independence:* Ole Aggo Markussen interview, Nuuk, Greenland, December 2023.

109   *"They're cheating us":* Lars Erik Gabrielsen interview, Ilulissat, Greenland, December 2023.

"THEN I'D BE THE ONE SENT TO GO FIND THEM"

113   *have been largely neutral in times of war:* Owen Greene, "Sweden: A History of Neutrality Ends After 200 Years," *The Conversation,* May 26, 2022, http://thecon versation.com/sweden-a-history-of-neutrality-ends-after-200-years-183583.

113   *applied for membership in NATO:* "Adapting Finnish Defence to NATO Membership," Helsinki Security Forum, August 24, 2023, https://helsinkisecurityforum

.fi/news/adapting-finnish-defence-to-nato-membership-and-changed-security
-situation.

113   *Turkey held up the process:* Ben Hubbard and Lara Jakes, "Turkey Back Sweden's NATO Bid," *New York Times,* January 23, 2024, https://www.nytimes
.com/2024/01/23/world/europe/turkey-sweden-nato.html.

114   *signs a Defense Cooperation Agreement:* "U.S. Signs Defense Cooperation
Agreement with Sweden," U.S. Department of State, press release, n.d.,
https://2021-2025.state.gov/u-s-signs-defense-cooperation-agreement
-with-sweden.

115   *spending by 34 percent: The Swedish Defence Commission's Report on the Development of the Military Defence,* April 26, 2024, https://www.regeringen.se
/contentassets/79646ada8654492993fe7108d95ac6d5/sammandrag-pa-en
gelska-av-starkt-forsvarsformaga-sverige-som-allierad-ds-20246.pdf.

115   *Swedish Lapland's 21-million-euro annual Christmas business:* "Lapland: The
World's Biggest Christmas Economy?," BBC Business Daily, December 18,
2024, https://www.bbc.co.uk/programmes/w3ct5zp1.

116   *suffered derailments on no fewer than:* Hilde-Gunn Bye, trans. Birgitte Hansen, "New Derailment on the Iron Ore Line Between Northern Sweden and
Norway," *High North News,* December 19, 2023, https://www.highnorth
news.com/en/new-derailment-iron-ore-line-between-northern-sweden
-and-norway.

116   *Sweden's NATO accession was expected:* G. Falco, N. Boschetti, and I. Nikas,
"Undercover Infrastructure Dual-Use Arctic Satellite Ground Stations,"
Center for International Governance Innovation, working paper, 2024, 291.

116   *arctic-space infrastructure with commercial broadband:* Atle Staalesen, "Liftoff for
the Norwegian Satellites That Will Provide Broadband to Circumpolar Arctic,"
Barents Observer, July 12, 2024, https://thebarentsobserver.com/en/2024/07/lift
off-norwegian-satellites-will-provide-broadband-circumpolar-arctic.

116   *after China broke ground on:* Melody Schreiber, "A New China-Iceland Arctic
Science Observatory Is Already Expanding Its Focus," *Arctic Today,* October 31,
2018, https://www.arctictoday.com/new-china-iceland-arctic-science-observa
tory-already-expanding-focus.

NOTES

116   *China Remote Sensing Satellite North Polar Ground Station:* "China Remote Sensing Satellite Ground Station," Aerospace Information Research Institute, Chinese Academy of Sciences, n.d., http://english.aircas.cn/research2020/bsi/202302/t20230213_327036.html.

118   *They need to keep moving:* Two interesting facts about Ernest Shackleton, the polar explorer: He kept expedition members busy by making tea, emboldening them to care for their team and encouraging a sense of value and purposefulness. Efforts, no matter how small, toward collective safety in a cold polar climate mattered deeply.

118   *Sweden's highest military commander long recognized:* Tim Martin, "Sweden 'Beefing Up' Military Presence in Arctic to Counter Russian Threat," *Breaking Defense,* October 4, 2023, https://breakingdefense.com/2023/10/sweden-beefing-up-military-presence-in-arctic-to-counter-russian-threat.

118   *a natural place to train:* Hilde-Gunn Bye, "Sweden's Chief of Defense Wants to Strengthen Military Presence in Northern Sweden," *High North News,* November 3, 2022, https://www.highnorthnews.com/en/swedens-chief-defense-wants-strengthen-military-presence-northern-sweden.

118   *In a private discussion among:* During the course of reporting and writing this book, I spoke with many current and former American military personnel who worked and trained in the arctic. I granted them anonymity to allow them to speak candidly about their experiences.

119   *"the 'pre-war era' has begun":* Hélène Bienvenu, Anne-Françoise Hivert, Chloé Hoorman, Philippe Jacqué, and Elise Vincent, "On NATO's Eastern Flank, on the Russian Border, the 'Pre-war Era' Has Begun," *Le Monde,* July 11, 2024, https://www.lemonde.fr/en/international/article/2024/07/11/on-nato-s-eastern-flank-on-the-russian-border-the-pre-war-era-has-begun_6680829_4.html.

120   *"didn't understand what's going on":* I heard similar anecdotes about American military capabilities in the arctic from several NATO partners.

121   *the largest cold-weather military exercise:* Steadfast Defender featured 90,000 service members from across the thirty-one NATO member states. It was met with hybrid attacks throughout the exercise, from January to June.

122   *resemble something out of the arctic:* Of all the reporting I undertook across the arctic, it was only while reporting on the front lines in Ukraine that my pen literally froze.

OPERATION TOUR DE HELSINKI

126   *"That's for sure—we are saying that openly":* Ben Taub, "Russia's Espionage War in the Arctic," *The New Yorker,* September 9, 2024, https://www.new yorker.com/magazine/2024/09/16/russias-espionage-war-in-the-arctic. With the stipulation that a source outside of this is hard to track, and that the *New Yorker* story itself had many inaccuracies and questionable observations that I witnessed the exact opposite of firsthand.

128   *Sámi could herd their reindeer:* Neil Kent, *The Sámi Peoples of the North: A Social and Cultural History* (C. Hurst, 2014).

129   *Vikings of Sweden and Norway once:* For additional reading on the Vikings, I suggest anything by historian Eleanor Rosamund Barraclough. One take-away from a review in the *New York Times* about one of Ms. Barraclough's books noted that referencing a "Viking era" would be akin to referring to today as the "Navy SEAL era."

130   *shots rang out across the border:* Thomas Nilsen, "Shots Were Fired by FSB, Wagner Defector Escaping to Norway Is Willing to Testify," *Barents Observer,* January 15, 2023, https://www.thebarentsobserver.com/security /shots-were-fired-by-fsb-wagner-defector-escaping-to-norway-is-willing -to-testify/163169.

131   *cyclists arrived first in Kirkenes:* Arwa Damon and Gul Tuysuz, "Above the Arctic Circle, Middle East Refugees Bike Their Way to Salvation," CNN, October 28, 2015, https://www.cnn.com/2015/10/28/europe/norway-russia -middle-east-refugees-bicycles/index.html.

131   *bomb shelters scattered from the:* Anne Kauranen, "Finland Counted Its Bomb Shelters and Found 50,500 of Them," Reuters, August 29, 2023, https://www .reuters.com/world/europe/finland-counted-its-bomb-shelters-found-50500 -them-2023-08-29.

132   *a veritable underground city:* Ines de la Cuetara, "Helsinki's 'Underground City' Reflects Tense Position as Russia's Neighbor," ABC News, May 12, 2022, https://abcnews.go.com/International/helsinkis-underground-city-re flects-tense-position-russias-neighbor/story?id=84668764.

132   *said that Finland's cooperation with:* "Comment by Foreign Ministry Spokes-woman Maria Zakharova on the Defence Cooperation Agreement between Finland and the US," Министерство Иностранных Дел Российской Федерации, December 19, 2023, https://www.mid.ru/ru/foreignpolicy/news /1921855/?lang=en.

133   *leads to the death of a:* Aleksi Teivainen, "Heavy Snowfall Causes Traffic Chaos in Helsinki Region," *Helsinki Times,* April 24, 2024, https://www.helsinkitimes .fi/finland/finland-news/domestic/25156-heavy-snowfall-causes-traffic-chaos -in-helsinki-region.html.

133   *they only seem to rise:* Atle Staalesen, "New Defence Agreement: Finland Invites American Troops to Bases in Lapland," *Barents Observer,* December 18, 2023, https://thebarentsobserver.com/en/security/2023/12/finland-invites-american -troops-far-northern-bases.

134   *"strengthen the border with Finland":* "ДШРГ Русич," *Telegram,* September 9, 2024 (in Russian), https://t.me/dshrg2/2268.

134   *to cross back into Russia:* Thomas Nilsen, Atle Staalesen, and Olesia Krivtsova, "Wagner Defector Detained by Norwegian Police in Attempted Illegal Bor-der Crossing Back to Russia," *Barents Observer,* September 22, 2023, https:// thebarentsobserver.com/en/2023/09/wagner-defector-detained-norwegian -police-attempted-illegal-border-crossing-back-russia.

134   *reported multiple airspace violations:* "Finland Suspects Four Russian Military Planes Violated Its Airspace," Reuters, June 14, 2024, https://www.reuters .com/world/europe/finland-suspects-four-russian-military-planes-violated -its-airspace-2024-06-14.

134   *a Russian submarine near Helsinki:* Arthur Herman, "Sweden and Finland Will Help NATO Counter Russia in the Arctic," *Wall Street Journal,* June 9, 2022, https://www.wsj.com/articles/sweden-and-finland-will-help-nato-confront -russia-in-the-arctic-ice-submarine-nuclear-11654786841.

NOTES

134   *like a now-abandoned proposal to construct a tunnel:* Finbarr Berningham, "Finnish Developers Push for Chinese-Built Tunnel to Estonia Despite Souring Mood Towards Beijing," *South China Morning Post,* March 17, 2024, https://www.scmp.com/news/china/diplomacy/article/3255653/fin nish-developers-push-chinese-built-tunnel-estonia-despite-souring -mood-towards-beijing.

134   *dependent on certain Russian imports:* Paso Peiponen, "Tehdas Suomessa jalostaa venäläistä nikkeliä ja palauttaa sotateollisuudelle yleistä kuparia— näin toimitusjohtaja kommentoi," *Yleisradio Oy,* updated April 4, 2024, https://yle.fi/a/74-20079921.

135   *construction of a fence planned:* Kostya Manenkov and Sergei Grits, "NATO Member Finland Breaks Ground on Russia Border Fence," AP News, April 15, 2023, https://apnews.com/article/finland-russia-nato-border-fence-illegal -migration-d89ca5f0105f604a0a8fd3523d8a7b00.

THE TYPHOON

138   *was born off the coast:* "Digital Typhoon: Typhoon 202213 (MERBOK). General Information (Pressure and Track Charts)," Kitamoto Asanobu, National Institute of Informatics, n.d., http://agora.ex.nii.ac.jp/digital-typhoon/sum mary/wnp/s/202213.html.en.

138   *industry providing 60 percent:* The Economic Value of Alaska's Seafood Industry, Alaska Seafood Marketing Institute, April 23, 2024, https://www.alaskaseafood .org/resource/economic-value-report-april-2024.

139   *miles in nearly every direction:* Derrick Bryson Taylor, "Western Alaska Lashed by Strongest Storm in Years," *New York Times,* September 16, 2022, https:// www.nytimes.com/2022/09/16/us/alaska-storm-typhoon-merbok.html.

139   *volleyball was canceled:* Diana Haecker, "Mega-Storm Hits Region, Causing Massive Destruction," *The Nome Nugget,* September 22, 2022, https://nomenug get.com/news/mega-storm-hits-region-causing-massive-destruction.

140   *were best friends from high school:* Maksim Teiuraut, interviews with author through a translator, August–October 2024. From more than six hours of

interviews with Maksim, documents obtained from the Department of Homeland Security, and in conversations with the legal representative for Max and Sergei, I was able to understand the scope and direness of their situation and their strenuous journey to American shores. In October 2024, Max was arrested by the Seattle Police Department on numerous charges, including assault and theft of a motor vehicle (Report Number 2024-289533). He was found in possession of narcotics thought to be crystal meth.

140   *from a military draft and:* "'Already Dead': Siberia's Homeless Pushed to Join Russia's War on Ukraine," Radio Free Europe, September 2, 2024, https://www.rferl.org/a/russia-ukraine-war-homeless-recruiting-casualties/33096675.html.

140   *the impacts of climate change:* Joshua Yaffe, "The Great Siberian Thaw," *The New Yorker,* January 10, 2022, https://www.newyorker.com/magazine/2022/01/17/the-great-siberian-thaw.

140   *Conscription may have seemed enticing:* Todd Prince, "Sweetening a Bitter Pill: Russia Offers Debt Breaks, Other Benefits to Entice Draftees," Radio Free Europe, September 28, 2022, https://www.rferl.org/a/russia-mobilization-incentives-draft-ukraine-war/32056473.html.

141   *four-meter-long* Sever *with:* Max had tried in vain for months to get his boat back, hoping he might strike a deal with the government so he could return to living aboard while awaiting asylum. Then he went silent.

141   *for a training exercise called:* The U.S. Northern Command has "Noble Defender" and "Polar Dagger" (SOCNORTH is nestled underneath that exercise).

141   *In a prescient report from:* James Morten, "Special Operations Forces and the Arctic: Meeting North America's 21st Century Security Needs," Center for Arctic Security and Resilience (CASR), University of Alaska Fairbanks, https://www.uaf.edu/casr/projects/arctic-sof-minds.php.

142   *had sought to establish more:* The Alaska Mapping Executive Committee "coordinates the modernization of critical geospatial data and mapping products for Alaska," but is still far from reaching its goals for 2030.

142   *were needed to predict and:* NOAA collected data from ten sensors during Typhoon Merbok, though not all the sensors were NOAA ones. The

agency has committed funding toward increasing geospatial assets that can help determine water levels. See "Ex-Typhoon Merbok Post-Storm Data Response," Alaska Geospatial Office, November 2, 2022, https://storymaps .arcgis.com/stories/ab19c80f9a644d3a9741e56cde5a41ab.

142   *percent of which has been mapped:* "The Interagency Working Group on Ocean and Coastal Mapping Announces Progress on Mapping U.S. Ocean, Coastal, and Great Lakes Waters," Office of Coast Survey, NOAA, March 28, 2025, https://nauticalcharts.noaa.gov/updates/the-interagency-working-group-on -ocean-and-coastal-mapping-announces-progress-on-mapping-u-s-ocean -coastal-and-great-lakes-waters.

143   *lifting and displacing homes from:* Haecker, "Mega-Storm Hits Region, Causing Massive Destruction."

143   *who lived in the community:* "Special Forces Parachuted over St. Lawrence Island on the Eve of the Storm," *The Nome Nugget,* September 29, 2022.

144   *making the embarkation and disembarkation:* Three independent sources confirmed the activities of the Navy SEALs and other Special Operations forces during the storm.

145   *recalled telling Sergei during their:* Though I was in contact with Sergei Nachaev, he did not participate in any interviews.

145   *a memorandum of Don Cossacks:* A. E. Janke, "The Don Cossacks on the Road to Independence," *Canadian Slavonic Papers, Revue Canadienne Des Slavistes* 12, no. 3 (Fall 1970): 273–94, https://www.jstor.org/stable/40866301.

A PERFECT STORM

147   *navy decided to reevaluate its:* "Strategic Outlook for the Arctic," U.S. Navy, January 2019, https://climateandsecurity.org/wp-content/uploads/2019/04 /strategic-outlook-for-the-arctic-jan-2019.pdf.

147   *told the Senate Armed Services Committee:* Megan Eckstein, "Navy to Release Arctic Strategy This Summer, Will Include Blue Water Arctic Operations," U.S. Naval Institute News, April 19, 2018, https://news.usni.org/2018/04/19/navy-to -release-arctic-strategy-this-summer-will-include-blue-water-arctic-operations.

147   *"the arctic triggered it"*: Eckstein, "Navy to Release Arctic Strategy This Summer: Will Include Blue Water Arctic Operations."

147   *that assessment and dozens of reports:* For one, there was the 2020 tri-service maritime strategy, *Advantage at Sea: Prevailing with Integrated All-Domain Naval Power*, December 2020, https://media.defense.gov/2020/dec/16/20 02553074/-1/-1/0/triservicestrategy.pdf.

147   *which send various ships and:* Meredith Chen, "China's Coastguard Takes Part in First Arctic Patrol with Russia," *South China Morning Post*, October 2, 2024, https://www.scmp.com/news/china/diplomacy/article/3280832/chinas-coast guard-takes-part-first-arctic-patrol-russia.

148   *and to pursue climate science:* In 2023, at a meeting of the Arctic Icebreaker Coordination Committee in Alexandria, Virginia, the coast guard announced it would cut back on how frequently it could support science missions, all but abandoning its congressional mandate to facilitate that research; this was a devastating blow to university scientists and researchers the world over.

148   *is building nuclear icebreakers for:* Finland is the world's preeminent designer and builder of icebreakers, with some 80 percent of the world's icebreakers connected in some way to Finnish construction firms. After Chinese vessels were spotted off the coast of Alaska, and with news that Russia was building "combat icebreakers," the United States, Finland, and Canada signed a joint agreement to "develop an implementation plan" for building "best-in-class Arctic and polar icebreakers" as part of a new ICE Pact. (See note for *It is one of four* starting on page 256.) This news came after years of hamstrung efforts by the coast guard to modernize its arctic fleet, which I touch on throughout the chapter. See Matthew Schilke, "Will Finland's Shipbuilding Expertise Break Diplomatic Ice with the US?," *Yleisradio Oy*, March 11, 2025, https://yle.fi/a/74-20148846.

149   *"The ocean is a beat":* The coast guard seaman shared personal email correspondence between himself and family and friends.

149   *encounter Chinese or Russian naval:* Matt Seyler and Tal Axelrod, "Russian and Chinese Ships Patrolled 'near Alaska' but Were Not 'a Threat,' US Officials Say," ABC News, August 6, 2023, https://abcnews.go.com/Politics/rus sian-chinese-ships-patrolled-alaska-threat-us-officials/story?id=102058344.

NOTES

149 *jets that enter Alaskan airspace:* "Two Russian Strategic Bombers Flew near Alaska," Reuters, February 7, 2024, https://www.reuters.com/world/two-rus sian-strategic-bombers-flew-near-alaska-2024-02-07.

150 *Purcell had deployed with:* Peter Purcell, multiple conversations with author between classes on base, January 2024, Kodiak, Alaska. Lieutenant Purcell also noted that the Polar Security Cutters would go some distance to improving the resources they do have. He said, "At any given time we have one boat" out in the Bering. "It's an impossible area to cover. There are massive blind spots all over. The air stations can't cover it. More ice-faring ships wouldn't solve the problem, but it would help."

150 *knew that the Polar Security:* "Coast Guard Acquisitions: Polar Security Cutter Needs to Stabilize Design Before Starting Construction and Improve Schedule Oversight," U.S. Government Accountability Office, press release, July 27, 2023, https://www.gao.gov/products/gao-23-105949.

150 *as ice-capable as the* Healy: There is, as this book goes to press, a third, smaller icebreaker, deployed to the Great Lakes.

150 *allocated funds for a proposed:* Trevor Hughes, "'Big Beautiful Bill' Supercharges Coast Guard's Arctic Icebreaker Fleet," *USA Today,* July 10, 2025, https://usatoday.com/story/news/politics/2025/07/10/trump-tax-bill-ice breakers-arctic-northwest-passage/84518867007.

151 *"Once we have the detailed design":* "Review of Fiscal Year 2024 Budget Request for the Coast Guard," Committee on Transportation and Infrastructure, April 18, 2023, https://transportation.house.gov/calendar/eventsingle .aspx?EventID=406261.

151 *Admiral Linda L. Fagan:* John Ismay, "Admiral Fagan Is Fired as Coast Guard Commandant" *New York Times,* January 21, 2025, https://www.nytimes.com /2025/01/21/us/politics/admiral-fagan-fired-coast-guard.html. In January 2025, Admiral Fagan was unceremoniously fired by President Trump, who cited, among other reasons, retention rates and the icebreaker acquisition program.

151 *the GAO found that discontinuity:* "Arctic Region: Factors That Facilitate and Hinder the Advancement of U.S. Priorities," U.S. Government Accountabil-

ity Office, press release, September 6, 2023, https://www.gao.gov/products /gao-23-106002.

153 *the night before our flight:* Steve Williams, "Coast Guard Rescues 4 Kodiak Mariners off Cape Chiniak," *Kodiak Daily Mirror,* January 24, 2024, https:// www.kodiakdailymirror.com/news/article_57ea1154-ba67-11ee-96f2-1f59 a5cd6f4c.html.

155 *by one standard to be:* As defined by the Arctic Research and Policy Act of 1984.

156 *a slurry of calcium magnesium:* This is a deicer alternative to road salt.

157 *when roughly 100 vessels transited:* Vessel traffic in U.S. Arctic waters could increase by another third by 2030. See "Less Ice, More Ships," Ocean Conservancy, information page, n.d., https://oceanconservancy.org/ls/shipping-ber ing-strait-region/overview.

157 *Alice Rogoff, publisher of:* The reported conversation took place during a state visit by Barack Obama in 2015. Rogoff and her Arctic Assembly founding partner, Ólafur Ragnar Grímsson, had for years expressed interest in opening up commercial activity and trading with China in the arctic. See Casey Kelly, "Could Western Alaska Become the New Panama Canal?," *KTOO,* April 11, 2013, http://www.ktoo.org/2013/04/11/could-western-alaska-be come-the-new-panama-canal.

157 *told President Barack Obama in:* Obama was the first sitting U.S. president to cross the Arctic Circle. See "In Alaska, Obama Becomes First President to Enter the Arctic," *MSNBC,* September 2, 2015, https://www.msnbc.com/msnbc/alas ka-obama-becomes-first-president-enter-the-arctic-msna674601.

158 *So successful was the* Rush: "Time Line 1700–1899," U.S. Coast Guard, information page, n.d., https://www.history.uscg.mil/Complete-Time-Line/Time -Line-1700-1800.

158 *returned most of them to:* Operation Clean Sweep occurred for a few years around this time. One report, the "Final Record of Decision" for the Bear Creek Radio Relay Station, recounts these efforts, though several other Operation Clean Sweep initiatives have taken place in different contexts and climates since.

158 *dump sites for naval and:* "ADAK Naval Air Station Site Profile," U.S. Environmental Protection Agency, information sheet, n.d., https://cumulis.epa.gov/supercpad/CurSites/csitinfo.cfm?id=1000128.

160 *in Juneau, said that year:* "Coast Guard Completes Arctic Shield 2013," *Marine Link,* November 4, 2013, https://www.marinelink.com/news/completes-arctic-shield360564.

160 *Waesche was similar in design:* Obendorf's case files and internal communications were culled from Laurie Powell, *Troubled Waters: The Legacy of USCG Hero BM3 Travis R. Obendorf ~ OBIE* (self-published, 2023).

163 *released a set of geographic:* "Announcement of U.S. Extended Continental Shelf Outer Limits," U.S. Department of State, press release, December 19, 2023, https://web.archive.org/web/20240109020824/https://www.state.gov/announcement-of-u-s-extended-continental-shelf-outer-limits.

164 *a majority of its domestic fisheries:* This is according to the Resource Development Council for Alaska. "Alaska's Fishing Industry," information page, n.d., https://www.akrdc.org/fisheries.

164 *federal lands that generated:* "Revenue and Disbursements from Oil and Natural Gas Production on Federal Lands," September 22, 2020, https://www.congress.gov/crs_external_products/R/PDF/R46537/R46537.2.pdf.

165 *known by its contemporary name:* Ray Hudson, "The Story of a Name: Unalaska or Dutch Harbor?," The City of Unalaska, Alaska, n.d., https://www.ciunalaska.ak.us/community/page/story-name-unalaska-or-dutch-harbor.

165 *next deep-water port to Nome:* Mark Thiessen, "Cruising to Nome: The First U.S. Deep Water Port for the Arctic to Host Cruise Ships, Military," AP News, June 18, 2023, https://apnews.com/article/alaska-arctic-port-nome-china-russia-588201b311513709404344fbc0d0e913.

TOOLIK

169 *per day during its lowest:* "Historic Throughput," Alyeska Pipeline, information sheet, n.d., https://www.alyeska-pipe.com/historic-throughput.

169   *began flooding back to:* "NSF's 10 Big Ideas—About NSF," National Science Foundation, information sheet, n.d., https://new.nsf.gov/about/history/big -ideas.

169   *Studies on arctic climate suggest:* Jacob Judah, "Russia's Warming Arctic Is a Climate Threat: War Has Shut Scientists Out of It," *New York Times,* October 22, 2024, https://www.nytimes.com/2024/10/22/climate/russia -alaska-arctic-global-warming.html.

169   *the amount of carbon in:* Kristina Kiest, "Permafrost and the Global Carbon Cycle," National Oceanic and Atmospheric Administration/Arctic, report, October 31, 2019, https://arctic.noaa.gov/report-card/report-card-2019/per mafrost-and-the-global-carbon-cycle.

169   *sitting atop more permafrost than:* "Thawing Permafrost Threatens Areas of NGA Interest," National Geospatial-Intelligence Agency, information page, n.d., https://web.archive.org/web/20250116111447/https://www.nga.mil /news/Thawing_Permafrost_Threatens_Areas_of_NGA_Interest.html.

172   *the only arctic tundra in:* Tundra can be either alpine habitats above the tree line or habitats beyond the northern tree line at sea level.

173   *a research editorial published in:* Ylva Sjöberg, Frédéric Bouchard, Susanna Gartler, Annett Bartsch, and Donatella Zona, "Focus on Arctic Change: Transdisciplinary Research and Communication," *Environmental Research Letters* 18, no. 1 (2023): 010201, https://doi.org/10.1088/1748-9326/acabd7.

173   *be those few degrees:* Darcy Frey, "George Divoky's Planet," *New York Times Magazine,* January 6, 2002, https://www.nytimes.com/2002/01/06/maga zine/george-divoky-s-planet.html.

173   *The rest of the world then followed:* There is, naturally, a heated debate over this as well. See James E. Overland, Baek-Min Kim and Yoshihiro Tachibana, "Communicating Arctic-Midlatitude Weather and Ecosystem Connections: Direct Observations and Sources of Intermittency," *Environmental Research Letters* 16, no. 10 (2021): 105006, https://doi.org/10.1088/1748-9326/ac25bc.

175   *an NSF–funded project to study:* "Freeze or Famine: Do Uncoupled Tempera- ture and Light Regimes Drive Unique Seasonal Production-Demand Rela- tionships in Arctic Spring-Stream Ecosystems?," Award #1947993, National

Science Foundation, June 12, 2020, https://www.nsf.gov/awardsearch/show Award?AWD_ID=1947993.

176   *"As an unintended consequence"*: Anurag Bisen, "India's G20 Presidency: Opportunity to Resume Engagement in the Arctic," Arctic Institute, Center for Circumpolar Security Studies, February 28, 2023, https://www.thearcticinstitute.org/india-g20-presidency-opportunity-resume-engagement-arctic.

176   *melting of glaciers in the Himalayan region:* This is known as the Third Pole Approach, which contends that research in the Himalayas can correspond to that in the arctic. Critics argue that there is little cooperation between the two research groups, and that India and other Asian countries like the United Arab Emirates are using this concept as a through tunnel into arctic decision-making.

177   *"China describes the Arctic"*: Rush Doshi, Alexis Dale-Huang, and Gasqi Zhang, "Northern Expedition: China's Arctic Activities and Ambitions," Brookings Institution, April 2021, https://www.brookings.edu/articles/northern-expedition-chinas-arctic-activities-and-ambitions.

177   *was planned to run from August:* Hanno Kinkel, "Postponement of ArcOP IODP Expedition #377," European Consortium for Ocean Research Drilling, April 7, 2022, https://www.ecord.org/postponement-of-arcop-iodp-expedition-377.

178   *Plastics pollution in the arctic:* Melanie Bergmann, France Collard, Joan Fabres, Geir Gabrielsen, Jennifer Provencher, et al., "Plastic Pollution in the Arctic," *Nature Reviews: Earth & Environment* 3, 323–37 (2022), https://www.nature.com/articles/s43017-022-00279-8.

DRONES OVER DELTA JUNCTION

181   *up to 50 miles per hour:* Adrian Peterson, "Military Report: Weather Delays Launch of Joint Pacific Multi-National Readiness Center Exercise," *Alaska's News Source: Fairbanks,* February 16, 2024, https://www.webcenterfairbanks.com/2024/02/17/military-report-weather-delays-launch-joint-pacific-multi-national-readiness-center-exercise.

183   *exercise with doctrine writers:* I was scheduled to speak with Captain Edward A. Garibay, Doctrine Author of Arctic Operations at the Combined Arms Doctrine Directorate, in October 2024. Captain Garibay canceled the interview days before, citing deadlines.

184   *almost 80 percent of the:* John E. Whitley, "Underfunding the Army Has Risky Implications," Association of United States Army, January 10, 2023, https://www.ausa.org/publications/underfunding-army-has-risky-implications.

185   *Even howitzers freeze:* Joseph Trevithick, "Army Faces Fight Just to Survive in the Arctic," *The War Zone,* October 17, 2023, https://www.twz.com/army-faces-fight-just-to-survive-in-the-arctic.

186   *detected four Russian Tu-series strategic:* Jon Haworth, "US Detects Russian Aircraft Flying in Alaska Air Defense Identification Zone," ABC News, February 7, 2024, https://abcnews.go.com/US/us-detects-russian-aircraft-flying-alaska-air-defense/story?id=107014947.

186   *"layered defense network of satellites":* "NORAD Detects and Tracks Russian Aircraft Operating in the Alaskan Air Defense Identification Zone," North American Aerospace Defense Command, press release, February 18, 2025, https://www.norad.mil/Newsroom/Press-Releases/Article/4070006/norad-detects-and-tracks-russian-aircraft-operating-in-the-alaskan-air-defense.

186   *In late January 2023:* Christina Wilkie and Emma Kinery, "U.S. Shoots Down Second 'High Altitude Object' Days After Chinese Spy Balloon," *CNBC,* February 10, 2023, https://www.cnbc.com/2023/02/10/us-shoots-down-second-high-altitude-object-on-bidens-orders.html; Jim Garamone, "F-22 Safely Shoots Down Chinese Spy Balloon off South Carolina Coast," US Department of Defense, February 4, 2023, https://www.defense.gov/News/News-Stories/Article/article/3288543/f-22-safely-shoots-down-chinese-spy-balloon-off-south-carolina-coast.

187   *home to more weather-balloon launch:* William J. Broad, "A Rising Awareness That Balloons Are Everywhere in Our Skies," *New York Times,* February 14, 2023, https://www.nytimes.com/2023/02/14/science/weather-balloons-stratosphere.html.

187 *"Especially in the arctic, the two most useful things"*: Pierre Leblanc, phone interview with author, September 2024.

187 *38.6 billion Canadian dollars through 2043: Our North, Strong and Free: A Renewed Vision for Canada's Defence,* Government of Canada, May 3, 2024, https://www.canada.ca/en/department-national-defence/corporate/reports-publications/north-strong-free-2024.html.

187 *first arctic base since:* David Ljunggren, "Canada Opposition Head Says He Will Slash Foreign Aid to Build Arctic Base," Reuters, February 2, 2025, https://www.reuters.com/world/americas/canada-opposition-head-says-he-will-slash-foreign-aid-build-arctic-base-2025-02-10.

188 *"develop and submit recommendations"*: FOIA Request, Department of Defense Inspector General, Ref: DODOIG-2020-000450, submitted January 27, 2020. Yielded DODIG-2020-051 Redacted. For full documents, go to https://kennethrrosen.com/arctic.

188 *section of which is fully:* For some reason, the footnotes to redacted sections of the report were not included when the Freedom of Information Act request was returned. Reading between the redacted sections led to the analysis made herein.

188 *flew over or near:* "Protection of Certain Facilities and Assets from Unmanned Aircraft," Section 130i, title 10, USC, of the report makes no distinction in size.

191 *An army handbook published in 2020:* Center for Army Lessons, review of *Mountain Warfare and Cold Weather Operations: Leader's Book,* U.S. Army, April 16, 2020, https://www.army.mil/article/247964/20_14_mountain_warfare_and_cold_weather_operations_leaders_book.

191 *wrote in* Science *magazine in:* H. Wexler, "Modifying Weather on a Large Scale," *Science* 128, no. 3331 (October 31, 1958): 1059–63, https://doi.org/10.1126/science.128.3331.1059.

191 *News of Dr. Wexler's article:* Walter Sullivan, "H-Bombs Visioned as Thawing Pole; But Scientist Warns That Arctic Blast Could Lead to a New Ice Age," *New York Times,* November 2, 1958, https://www.nytimes.com/1958/11/02/archives/hbombs-visioned-as-thawing-pole-but-scientist-warns-that-arctic.html.

NOTES

192   *A senior defense intelligence:* Author conversation with unnamed active-duty military personnel, 2023.

ANGELS IN THE DARK

194   *Delilah's been a funeral director:* Despite its large geographic size, Alaska is a small community, so I've used a pseudonym to protect Delilah's privacy and career. Conversations with author, 2024.

197   *one soldier was quoted as:* Robert D. Kaplan, *Hog Pilots, Blue Water Grunts: The American Military in the Air, at Sea, and on the Ground* (Random House, 2007), 36–38.

198   *one specialist tells me through:* I'd met a number of relatives and friends of military personnel, as well as service members themselves, while lecturing and traveling around Fairbanks. They asked not to be identified in order to candidly discuss their experiences in the military.

198   *523 service members in the army:* "Department of Defense Releases Its Annual Report on Suicide in the Military for Calendar Year 2023," U.S. Department of Defense, press release, November 14, 2024, https://www .defense.gov/News/Releases/Release/Article/3964785/department-of-de fense-releases-its-annual-report-on-suicide-in-the-military.

198   *are more likely to die:* Will Huntsberry, "Active-Duty Service Members Are Dying by Suicide at Alarming Rates," *Voice of San Diego,* n.d., https://voice ofsandiego.org/military-suicide-investigation.

198   *reporting seventeen suicides among its:* Dan Bross, "Alaska Army Suicides Drop as Leaders Push Programs to Improve Soldiers' Lives," Alaska Public Media, January 10, 2023, https://alaskapublic.org/news/2023-01-10/alaska -army-suicides-drop-as-leaders-push-programs-to-improve-soldiers-lives.

198   *destigmatize mental health care:* Rachel Cassandra, "Alaska's Army Division Is Combatting High Suicide Rates with Mandatory Wellness Counseling," Alaska Public Media, July 5, 2023, https://alaskapublic.org/2023/07/05/alas kas-army-division-is-combatting-high-suicide-rates-with-mandatory-well ness-counseling.

NOTES

199 *four soldiers died in suspected:* Wyatt Olson, "Army Alarmed by Spate of Suspected Soldier Suicides in Alaska," *Stars and Stripes,* November 9, 2022, https://www.stripes.com/theaters/us/2022-11-08/army-alaska-soldier-sui cides-7980017.html.

199 *Drawing from the Qungasvik model:* Brandon Kapelow, "Native-Led Suicide Prevention Program Focuses on Building Community Strengths," NPR, September 10, 2024, https://www.npr.org/2024/09/10/nx-s1-5100913/sui cide-prevention-program-alaska-native-community-mental-health.

200 *may mitigate the risk of:* Benjamin Trachik, Emma H. Moscardini, Michelle L. Ganulin, Jen L. McDonald, Ashlee B. McKeon, et al., "Perceptions of Purpose, Cohesion, and Military Leadership: A Path Analysis of Potential Primary Prevention Targets to Mitigate Suicidal Ideation," *Military Psychology* 34, no. 3 (2021): 366–75, https://doi.org/10.1080/08995605.2021.1962184.

200 *three-hour course begins with:* This information is based on a review of the slides prepared for NCOs that were given to me by Dr. James Morton and his team.

THE LAST TREE

205 *largely in Siberia and Canada:* Matthew Jones, Sander Veraverbeke, Niels Andela, Stefan Doerr, Crystal Kolden, et al., "Global Rise in Forest Fire Emissions Linked to Climate Change in the Extratropics," *Science* 386, no. 6719 (2024), https://www.science.org/doi/10.1126/science.adl5889.

205 *the size of West Virginia:* Or, for international readers, the size of the Netherlands.

205 *faster than they can regenerate:* Manuela Andreoni and Bryan Denton, "Parts of Canada's Boreal Forest Are Burning Faster Than They Can Regrow," *New York Times,* August 13, 2024, https://www.nytimes.com/interactive/2024/08/12/cli mate/canada-wildfires.html.

206 *world-class resources of copper:* "Ambler Access Update: House Resources Committee," Alaska Industrial Development and Export Authority, presentation, April 6, 2018, https://web.archive.org/web/20220812183723/https://www.akleg.gov/basis/get_documents.asp?session=30&docid=54616.

206   *"We deserve the same right to economic prosperity"*: "Alaska Business Leaders: Federal Regulations in Natural Petroleum Reserve Deal Major Setback to State," Alaska Oil & Gas Association, April 22, 2024, https://www.aoga.org /news/alaska-business-leaders-federal-regulations-in-natural-petroleum-re serve-deal-major-setback-to-state.

206   *"that caribou were heard over cash"*: Kavitha George, "Biden Administration Blocks Ambler Road, Strengthens Protections for NPR-A," Alaska Public Media, April 19, 2024, https://alaskapublic.org/2024/04/19/biden-adminis tration-blocks-ambler-road-strengthens-protections-for-npr-a. Ultimately, cash was louder. In September 2025, the Republican-controlled House voted to allow the resumption of the Ambler Access Project, greenlighting the construction of 211 miles of roadway that would cut through pristine and fragile Alaskan wilderness. Maxine Joselow, "House Votes to Advance a Mining Road Through the Alaskan Wilderness," *New York Times*, September 3, 2025, https://www.nytimes.com/2025/09/03/climate/ambler-access-project-road -alaska.html.

207   *rich deposits of lithium*: Rose Ragsdale, "Lithium Ignites Exploration Rush North," *Mining News,* August 24, 2023, https://www.miningnewsnorth.com /story/2023/08/04/news/lithium-ignites-exploration-rush-north/8037.html.

207   *U.S. military also needs antimony*: Max Graham, "America Needs Antimony for Weapons and Solar Panels. The Mining Industry Is Looking to Alaska," *Northern Journal,* November 15, 2024, https://www.northernjournal.com /america-needs-antimony-for-weapons-and-solar-panels-the-mining-in dustry-is-looking-to-alaska.

207   *165 feet and grew six feet around*: A magazine article by John Vaillant, "The Golden Bough," appeared in *The New Yorker* in 2002 and led to his writing *The Golden Spruce: A True Story of Myth, Madness, and Greed* (W. W. Norton, 2006).

209   *a white spruce*: Ben Weissenbach, "A Journey to the Northernmost Tree in Alaska," *Smithsonian,* June 1, 2021, https://www.smithsonianmag.com/sci ence-nature/journey-northernmost-tree-alaska-180977862. *Picea glauca* was discovered during this expedition.

210   *is a gluttonous baby with:* This story comes from Maria Bach Kreutzmann, ed., *Bestiarium Greenlandica: A Compendium of the Mythical Creatures, Spirits, and Strange Beings of Greenland* (Eye of Newt Books, 2022), though the interpretation and retelling are my own.

APPENDIX: REINING IN THE ARCTIC

212   *White House issued a National Strategy:* "Fact Sheet: The United States' National Strategy for the Arctic Region," The White House, October 7, 2022, https://bidenwhitehouse.archives.gov/briefing-room/statements-releases/2022/10/07/fact-sheet-the-united-states-national-strategy-for-the-arctic-region.

213   *In response to these challenges:* "Russia and China Eye NATO's 'Arctic Achilles Heel,'" France 24 News, June 23, 2022, https://www.france24.com/en/live-news/20220623-russia-and-china-eye-nato-s-arctic-achilles-heel.

214   *Though the "Arctic 8":* Duncan Depledge and Mathieu Boulègue, "New Military Security Architecture Needed in the Arctic," *Chatham House,* May 19, 2022, https://www.chathamhouse.org/2021/05/new-military-security-architecture-needed-arctic.

214   *The report identified a number:* Office of Under Secretary for Defense (Policy), *Report to Congress on Arctic Operations and the Northwest Passage,* Department of Defense, May 2011.

214   *The White House should create:* Marisol Maddox and Lyston Lea, "The Intelligence Community Must Evolve to Meet the Reality of Arctic Change," *Polar Perspectives* (Wilson Center), no. 13 (2023).

215   *There have been "dozens of incidents":* Adrian Peterson, "Military Report: Sullivan Calls on Navy to Reopen Base on Adak Island," *Alaska's News Source: Fairbanks,* October 9, 2024, https://www.webcenterfairbanks.com/2024/10/09/military-report-sullivan-calls-navy-reopen-base-adak-island.

215   *"Arctic nations' search-and-rescue":* David Palmer, "A Dose of Reality: Search and Rescue for Arctic Sustainable Development," *IntSights,* July 11, 2023, https://intsights.luminint.net/arctic-search-and-rescue.

215 *Washington should adapt this law:* Palmer, "A Dose of Reality."

216 *It's only natural that similar assets:* Tech Sergeant Jamie Spaulding, "New York Air Guard's 109th Airlift Wing Tests New LC-130 Engine," National Guard, October 20, 2022, https://www.nationalguard.mil/News/Article/31 94806/new-york-air-guards-109th-airlift-wing-tests-new-lc-130-engine.

216 *Fluctuating environmental conditions:* Kelsey A. Frazier, "Arctic Insecurity: The Implications of Climate Change for US National Security," *Journal of Indo-Pacific Affairs*, July 1, 2024.

216 *In February 2025, President Trump:* Jonathan Lambert, "National Science Foundation Fires Roughly 10% of Its Workforce," NPR, February 18, 2025.

216 *The White House must follow through:* Lord Henry Ashton, "Our Friends in the North: UK Strategy Towards the Arctic," *High North News,* December 8, 2023, https://www.highnorthnews.com/en/our-friends-north -uk-strategy-towards-arctic.

216 *By advocating for carbon neutrality:* Ashton, "Our Friends in the North."

216 *Washington should advocate for:* Ian Williams, Heather A. Conley, Nikos Tsafos, and Matthew Melino, "America's Arctic Moment: Great Power Competition in the Arctic to 2050," Center for Strategic & International Studies, March 30, 2020, https://www.csis.org/analysis/americas-arctic-moment-great-power-competi tion-arctic-2050.

216 *Because the private sector handles:* Ashton, "Our Friends in the North."

216 *By ratifying the United Nations Convention on the Law of the Sea:* Kenneth R. Rosen, "Opinion: A Growing Rivalry in the Arctic? Talk About a Cold War," *Washington Post,* July 6, 2023, https://www.washingtonpost.com/opinions /2023/07/06/arctic-cold-war-rivalry.

217 *If we want to maintain:* David Balton, Rosalie Debenham, Larry Hinzman, and Jane Lubchenko, "Policy and Science in Service to the Arctic and Its People: A Review of Arctic Policy from 2020 to 2024," *Polar Perspectives* (Wilson Center), no. 17 (2025).

217 *Intelligence strategies are important:* Maddox and Lea, "The Intelligence Community Must Evolve to Meet the Reality of Arctic Change."

NOTES

217   *By sharing intelligence and developing:* Ashton, "Our Friends in the North."

218   *If Washington doesn't demonstrate:* Jeffrey Gettleman and Ivor Prickett, "Greenland's Big Moment," *New York Times,* February 20, 2025, https://www.nytimes.com/2025/02/20/world/europe/greenland-trump-denmark.html.

218   *Washington should back climate initiatives:* Damian Carrington, "Rock 'Flour' from Greenland Can Capture Significant CO2, Study Shows," *The Guardian,* May 30, 2023, https://www.theguardian.com/environment/2023/may/30/rock-flour-greenland-capture-significant-co2-study.

218   *Supporting this project in particular:* Yasmin Tayag, "The Surreal Abundance of Alaska's Permafrost Farms," *The New Yorker,* August 30, 2022, https://www.newyorker.com/news/annals-of-a-warming-planet/the-surreal-abundance-of-alaskas-permafrost-farms.

SELECTED BIBLIOGRAPHY

Albanov, Valerian. *In the Land of White Death: An Epic Story of Survival in the Siberian Arctic.* Translated by Alison Anderson. Introduction by David Roberts. Preface by Jon Krakauer. Random House, 2001.

Amor, Norman. *Beyond the Arctic Circle.* University of British Columbia Press, 1992.

Andrée, S. A., Nils Strindberg, and K. Frænkel. *Andrée's Story.* Translated by Edward Adams-Ray. Viking Press, 1930.

Applebaum, Anne. *Gulag: A History of the Soviet Camps.* Doubleday, 2003.

Atwood, Margaret. *Strange Things: The Malevolent North in Canadian Literature.* Oxford University Press, 1996.

Barrett, Andrea. *The Voyage of the Narwhal.* W. W. Norton, 1999.

Bierman, Paul. *When the Ice Is Gone: What a Greenland Ice Core Reveals About Earth's Tumultuous History and Perilous Future.* W. W. Norton, 2024.

Bjarnason, Egill. *How Iceland Changed the World: The Big History of a Small Island.* Penguin, 2021.

Breum, Martin. *Cold Rush: The Astonishing True Story of the New Quest for the Polar North.* Bloomsbury, 2018.

Breum, Martin. *The Greenland Dilemma.* Translated by Kevin McGwin. Royal Danish Defense College, 2015.

Broberg, Henrik. *Da byen ble stille.* Margbok, 2014.

Brown, Stephen. *The Last Viking: The Life of Roald Amundsen.* Douglas & McIntyre, 2012.

Buchanan, Elizabeth. *Red Arctic: Russian Strategy Under Putin.* Brookings Institution Press, 2023.

SELECTED BIBLIOGRAPHY

Burke, Danita Catherine. *Diplomacy and the Arctic Council*. McGill-Queen's University Press, 2019.

Burns, James R. *The Cold Coasts*. Memini, 2006.

Byers, Michael. *Who Owns the Arctic? Understanding Sovereignty Disputes in the North*. Douglas & McIntyre, 2010.

Carson, Rachel. *The Sea Around Us*. Oxford University Press, 1951.

Chekhov, Anton Pavlovich. *The Island: A Journey to Sakhalin*. Century, 1987.

Chester, Sharon. *The Arctic Guide: Wildlife of the Far North*. Princeton University Press, 2016.

Cone, Marla. *Silent Snow: The Slow Poisoning of the Arctic*. Grove Press, 2006.

Davies, Thomas D. *Robert E. Peary at the North Pole: A Report to the National Geographic Society by the Foundation for the Promotion of the Art of Navigation*. Edited by Eloise E. Davies. Starpath Publications, 2009.

Egan, Dan. *The Death and Life of the Great Lakes*. W. W. Norton, 2017.

Ehrlich, Gretel. *This Cold Heaven: Seven Seasons in Greenland*. Vintage, 2003.

Eiricksson, Leifur. *The Vinland Sagas: The Icelandic Sagas about the First Documented Voyages Across the North Atlantic*. Penguin Classics, 2008.

Emmerson, Charles. *The Future History of the Arctic*. PublicAffairs, 2010.

Fossett, Renee. *In Order to Live Untroubled: Inuit of the Central Arctic, 1550 to 1940*. University of Manitoba Press, 2001.

Frison-Roche, Roger. *Hunters of the Arctic*. J. M. Dent, 1974.

Gaddis, John Lewis. *The Cold War: A New History*. Penguin, 2005.

Gertner, Jon. *The Ice at the End of the World: An Epic Journey into Greenland's Buried Past and Our Perilous Future*. Random House, 2019.

Gjørv, Gunhild Hoogensen, Marc Lanteigne, and Horatio Sam-Aggrey, eds. *Routledge Handbook of Arctic Security*. Taylor & Francis Group, 2020.

Goodman, Sherri. *Threat Multiplier: Climate, Military Leadership, and the Fight for Global Security*. Island Press, 2024.

Greene, A. Kendra. *The Museum of Whales You Will Never See: And Other Excursions to Iceland's Most Unusual Museums*. Penguin, 2020.

Griffiths, Franklin, ed. *The Politics of the Northwest Passage*. McGill-Queen's University Press, 1987.

SELECTED BIBLIOGRAPHY

Guldager, Ole. *Americans in Greenland in World War Two*. Arctic Sun, 2019.

Guldager, Ole. *The History of Narsarsuaq*. Vol. 1 of *Greenland History* series. Arctic Sun, 2016.

Hartman, Darrell. *Battle of Ink and Ice: A Sensational Story of News Barons, North Pole Explorers, and the Making of Modern Media*. Viking, 2023.

Hays, Otis, Jr. *Alaska's Hidden Wars: Secret Campaigns on the North Pacific Rim*. University of Alaska Press, 2004.

Hoffman, David E. *The Dead Hand: The Untold Story of the Cold War Arms Race and Its Dangerous Legacy*. Knopf Doubleday, 2009.

Kaplan, Robert D. *The Revenge of Geography: What the Map Tells Us About Coming Conflicts and the Battle Against Fate*. Random House, 2013.

Kaplan, Robert D. *Monsoon: The Indian Ocean and the Future of American Power*. Random House, 2010.

Kaplan, Robert D. *Hog Pilots, Blue Water Grunts: The American Military in the Air, at Sea, and on the Ground*. Vintage Departures, 2008.

Kaplan, Robert D. *The Coming Anarchy: Shattering the Dreams of the Post Cold War*. Knopf Doubleday, 2001.

Kaplan, Robert D. *An Empire Wilderness: Traveling into America's Future*. Vintage Departures, 1999.

Kauffman, John. *Alaska's Brooks Range: The Ultimate Mountains*. Mountaineers Books, 1992.

Kent, Neil. *The Sámi Peoples of the North: A Social and Cultural History*. C. Hurst, 2014.

Kent, Neil. *The Soul of the North: A Social, Architectural and Cultural History of the Nordic Countries, 1700–1940*. Reaktion Books, 2004.

Kersting, Rudolf. *The White World: Life and Adventures Within the Arctic Circle Portrayed by Famous Living Explorers*. Reprint, Forgotten Books, 2018.

Koch, Niels Elers. *Trap Grønland*. Trap Denmark, 2022.

Kolbert, Elizabeth, ed. *Arctic: An Anthology*. Granta Books, 2008.

Kolbert, Elizabeth. *Field Notes from a Catastrophe: Man, Nature, and Climate Change*. Bloomsbury, 2006.

Kpomassie, Tété-Michel. *Michel the Giant: An African in Greenland*. Translated by James Kirkup and Ros Schwartz. Penguin, 2022.

SELECTED BIBLIOGRAPHY

Krakauer, Jon, and David Roberts. *Iceland: Land of the Sagas*. Villard, 1998.

Krakauer, Jon. *Into the Wild*. Villard, 1996.

Kraska, James, ed. *Arctic Security in an Age of Climate Change*. Cambridge University Press, 2011.

Kreutzmann, Maria Bach, ed. *Bestiarium Greenlandica: A Compendium of the Mythical Creatures, Spirits, and Strange Beings of Greenland*. Eye of Newt Books, 2022.

Laxness, Halldór. *Independent People: An Epic*. Translated by J. A. Thompson. Alfred A. Knopf, 1946.

Lidegaard, Bo. *Uden mandat: En biografi om Henrik Kauffmann*. Gyldendal, 2020.

Lidegaard, Bo. *Defiant Diplomacy: Henrik Kauffmann, Denmark, and the United States in World War II and the Cold War, 1939–1958*. Translated by W. Gyl Jones. International Academic Publishers, 2003.

Loomis, Chauncey. *Weird and Tragic Shores: The Story of Charles Francis Hall, Explorer*. Edited by Jon Krakauer. Introduction by Andrea Barrett. Modern Library, 2000.

Lopez, Barry. *Arctic Dreams: Imagination and Desire in a Northern Landscape*. Charles Scribner's Sons, 1986.

Lutz, Birgit. *Nachruf auf die Arktis*. BTB Verlag, 2022.

Lyall, Ernie. *An Arctic Man: The Classic Account of Sixty-Five Years in Canada's North*. Formac, 1979.

Lynge, Aqqaluk. *Inuit: The Story of the Inuit Circumpolar Conference*. Atuakkiorfik, 1993.

Malaurie, Jean. *The Last Kings of Thule*. Translated by Adrienne Foulke. E. P. Dutton, 1982.

Mankiller, Wilma. *Every Day Is a Good Day: Reflections by Contemporary Indigenous Women*. Introduction by Gloria Steinem. Fulcrum, 2011.

McGrath, Melanie. *The Long Exile: A True Story of Deception and Survival in the Canadian Arctic*. Harper Perennial, 2007.

McPhee, John. *Coming into the Country*. Farrar, Straus and Giroux, 1977.

Morrison, Colin A. *Voyage into the Unknown: The Search for and Recovery of Cosmos 954*. Canada's Wings, 1982.

Muir, John. *Travels in Alaska*. Reprint, Dover, 2017.

SELECTED BIBLIOGRAPHY

Mundy, Robyn. *Cold Coast*. Ultimo Press, 2022.

Murkowski, Lisa. *Far from Home: An Alaskan Senator Faces the Extreme Climate of Washington, D.C.* Forum Books, 2025.

Nijhuis, Michelle. *Beloved Beasts: Fighting for Life in an Age of Extinction*. W. W. Norton, 2021.

O'Donoghue, Brian Patrick. *The Fairbanks Four: Murder, Injustice, and the Birth of a Movement*. Sourcebooks, 2025.

O'Neill, Dan. *The Firecracker Boys: H-Bombs, Inupiat Eskimos, and the Roots of the Environmental Movement*. St. Martin's Press, 1994.

Oxenhorn, Harvey. *Tuning the Rig: A Journey to the Arctic*. Zoland Books, 2000.

Petrow, Richard. *Across the Top of Russia: The Cruise of the USCGC Northwind into the Polar Seas North of Siberia*. David McKay, 1967.

Poelzer, Greg, William R. Morrion, P. Whitney Lackenbauer, and Ken S. Coates. *Arctic Front: Defending Canada in the Far North*. Thomas Allen, 2008.

Potter, Russell Alan. *Arctic Spectacles: The Frozen North in Visual Culture, 1818–1875*. University of Washington Press, 2007.

Proenneke, Richard, and Sam Keith. *One Man's Wilderness: An Alaskan Odyssey*. Alaska Northwest Books, 1999.

Rawlence, Ben. *The Treeline: The Last Forest and the Future of Life on Earth*. St. Martin's Press, 2022.

Reiss, Bob. *The Eskimo and the Oil Man: The Battle at the Top of the World for America's Future*. Business Plus, 2012.

Ritter, Christiane. *Woman in the Polar Night*. Translated by Jane Degras. Hamilton, 1956.

Rush, Elizabeth. *The Quickening: Creation and Community at the Ends of the Earth*. Milkweed Editions, 2023.

Rush, Elizabeth. *Rising: Dispatches from the New American Shore*. Milkweed Editions, 2018.

Sarotte, M. E. *Not One Inch: America, Russia, and the Making of Post–Cold War Stalemate*. Yale University Press, 2021.

Schmidt, Mikkelsen Peter. *North-East Greenland 1908–60: The Trapper Era*. Scott Polar Research Institute, 2008.

Sejersen, Frank. *Rethinking Greenland and the Arctic in the Era of Climate Change*. Taylor & Francis, 2015.

Sokolíčková, Zdenka. *The Paradox of Svalbard: Climate Change and Globalisation in the Arctic*. Foreword by Thomas Hylland Eriksen. Pluto Press, 2023.

Spohr, Kristina, and Daniel S. Hamilton, eds. *The Arctic and World Order*. Brookings Institution Press, 2020.

Stevenson, Lisa. *Life Beside Itself: Imagining Care in the Canadian Arctic*. University of California Press, 2014.

Stratigakos, Despina. *Hitler's Northern Utopia: Building the New Order in Occupied Norway*. Princeton University Press, 2020.

Strøksnes, Morten. *Shark Drunk: The Art of Catching a Large Shark from a Tiny Rubber Dinghy in a Big Ocean*. Translated by Tiina Nunnally. Jonathan Cape, 2017.

Tagaq, Tanya, and Jaime Hernandez. *Split Tooth*. Viking Press, 2018.

Vaillant, John. *The Golden Spruce: A True Story of Myth, Madness, and Greed*. W. W. Norton, 2006.

Vann, David. *Legend of a Suicide*. University of Massachusetts Press, 2008.

Veer, Gerrit de. *The Three Voyages of William Barents to the Arctic Regions: 1594, 1595 and 1596*. Edited by Laurens Koolemans Beynen. Hakluyt Society, 1853.

Vesaas, Tarjei. *The Ice Palace*. Gyldendal Norsk Forlag, 1963.

Vince, Gaia. *Nomad Century: How to Survive the Climate Upheaval*. Penguin, 2022.

Wallance, Gregory J. *Into Siberia: George Kennan's Epic Journey Through the Brutal, Frozen Heart of Russia*. St. Martin's Press, 2023.

Wheeler, Sara. *The Magnetic North: Notes from the Arctic Circle*. Farrar, Straus and Giroux, 2011.

Wilkins, Joe. *Gates of the Arctic National Park: Twelve Years of Wilderness Exploration*. Brown Books, 2018.

Williams, Glyn. *Arctic Labyrinth: The Quest for the Northwest Passage*. Penguin, 2009.

Winchester, Simon. *Pacific: Silicon Chips and Surfboards, Coral Reefs and Atom Bombs, Brutal Dictators, Fading Empires*. HarperCollins, 2015.

SELECTED BIBLIOGRAPHY

Woldstad, Wanny. *Wanny Get Your Gun: Wanny Woldstad—The First Woman Trapper in Svalbard*. Vågemot miniforlag, 2018.

Ylvisåker, Lina Nagell. *My World Is Melting: Living with Climate Change in Svalbard*. Translated by Kelsey Camacho. Yirisaaker AS, 2022.

Ytreland, Ivar. *A Trapper in North-East Greenland: Tales of a Forgotten Way of Life*. Translated by David Matthews. Steading Workshop, 2014.

INDEX

INDEX

INDEX

INDEX

INDEX

INDEX

National Ecological Observatory Network, 174
National Strategy for the Arctic Region (United
    States, 2022), 41, 63–64, 65, 212
NATO (North Atlantic Treaty Organization)
    borders and, 132–33, 134
    collective approach and, 38
    defense preparation, 18, 40, 41, 42, 115, 119,
        213, 241n
    Denmark and, 108
    espionage and, 95
    Finland and, 113, 132–33, 134, 213
    Greenland and, 111
    horizontal escalation and, 12
    Norway and, 78
    Sweden and, 113, 115, 116, 119, 134
    Ukraine war and, 115, 241n
    U.S. arctic presence and, 41–42, 61
natural resources
    Alaska, 39, 49, 50, 206–7, 240n
    Arctic Council and, 39
    Arctic Ocean, 73
    Canada, 207
    China, 207
    climate change and, 38, 163–64
    Cold War and, 9
    collective approach and, 10, 39
    vs. conservation, 206–7, 208–9
    Greenland, 93, 108, 218
    increased exploitation of, 11, 29
    Norway, 39, 240n
    policy recommendations, 218
    Russian arctic presence and, 47, 51, 52, 55, 93
    Svalbard, 52, 55
    Sweden, 93, 115
    territorial claims and, 6
    trade routes and, 74, 75
    war possibilities and, 93
Navalny, Aleksei, 242n
near-arctic state status, 9, 39, 54
Neqikitsuliaq, 210
Nielsen, Rasmus Leander, 108
Nikel, 130
Nome, Alaska, 64, 65, 165
NORAD, 64, 186
    gamesmanship and, 13
    U.S.–Canada North American Defense
        Command, 37
Nordsletten, Øyvind, 29
Norland, Richard B., 94
North (Arctic) Slope, 172
    See also Toolik Field Station
North American Aerospace Defense Command.
    See NORAD

North American Arctic. See Alaska; Canada; U.S.
    Coast Guard Bering Sea patrols
Northeast Passage (Northern Sea Route), 74, 75, 79
Northern Forum, 43
northern lights, 44, 195
northern poles, 225n
Northern Sea Route (NSR), 48
North Korea, Russian arctic presence and, 56
North Warning System (NWS), 40, 186, 187
Northwest Passage, 74, 81, 128, 216–17
Norway
    Arctic Ocean patrols, 71–73, 75–76, 77–78, 79–87
    borders and, 125–28, 129–30
    on collective approach, 38, 78–79, 240n
    defense preparations, 40
    experiences of, 18–19
    icebreakers and, 239n
    migrants and, 131–32
    NATO and, 78
    natural resources, 39, 240n
    Russian cooperation and, 78–79, 87
    security issues and, 37, 54
    Svalbard and, 22, 25–26, 28–29, 232n
    U.S. arctic presence and, 78, 96–97
    U.S. espionage and, 94
nuclear submarines, 37
Ny-Ålesund, 26

Obama, Barack, and administration, 97, 157, 273n
Obendorf, Travis R., 161
offshore oil drilling, 6, 209
    See also natural resources
Ohthere of Hålogaland, 128
oil. See natural resources
Olsvig, Sara, 43
Operation Arctic Shield, 159–61
Operation Clean Sweep, 273n
Operation NANOOK, 182
Ostebo, Thomas, 160
Ottawa Declaration, 214

Papp, Robert, Jr., 97
Parry, William Edward, 78, 255n
Pasvik River, 125–28
Patkotak, Josiah, 206
Peary, Robert, 60
permafrost, 11, 61, 62, 65, 169, 172
permafrost tunnel project, 192
Peter the Great (czar of Russia), 129
Philberth, Bernhard, 250n
Philberth, Karl, 250n
Pincus, Rebecca, 42
pipelines, 169, 170, 206

INDEX

INDEX

INDEX

INDEX

Russian and American Defense Assets in the Circumpolar North

□ Russian   ▲ American

Baffin Bay

GREENLAND

ATLANTIC OCEAN

Greenland Sea

ARCTIC

North Pole

SVALBARD
(Norway)

NORWAY

SWEDEN

FINLAND

FRANZ JOSEF LAND
(Russia)

Barents
Sea

Kara Sea

R U S S I A